Europe
A Geographical Survey of the Continent

Also by Roy E.H. Mellor

Geography of the USSR
Comecon: Challenge to the West
Eastern Europe: A Geography of the Comecon Countries

Europe

A Geographical Survey of the Continent

Roy E. H. Mellor and E. Alistair Smith

First published 1979 by
THE MACMILLAN PRESS LTD
London and Basingstoke
Associated companies in Delhi Dublin
Hong Kong Johannesburg Lagos Melbourne
New York Singapore and Tokyo

Photoset and printed
in Great Britain by
REDWOOD BURN LIMITED
Trowbridge and Esher

British Library Cataloguing in Publication Data

Mellor, Roy Egerton Henderson
Europe.
1. Europe – Description and travel – 1971 –
I. Title   II. Smith, E. Alistair
914   D923

ISBN 0–333–19250–8
ISBN 0–333–19251–6   Pbk

Contents

List of Figures

List of Tables

Preface

To understand Europe properly is best achieved by the study of its geography, not country by country, but as a large continental mass, seeking its unity within its diversity. Too often the emphasis seems to be laid upon Europe's divisions. Such a study, we firmly believe, should have a strong historical component, because Europe's landscapes and their patterns, the *mirror* of human impact over long periods, are the product of changes in man's response through time to his environment. The historical approach is not only useful to appreciate patterns of our own contemporary setting but essential if we are to venture to look forwards to try to perceive the Europe we may expect a quarter of a century from now. Our interpretation, in this volume, of Europe 2000 is one of many that may be modelled from a bewildering array of imponderables: we hope it will serve as a basis for debate and conjecture, and in the end prove not too far from reality.

Whatever the past connections, the Soviet Union has become so distinctive in its Eurasian setting that we feel it is no longer real to conceive 'European Russia' as distinct from 'Asiatic Russia' and consequently this vast country is best excluded from an overall view of Europe, except in passing reference relevant to its links with Eastern Europe through *Comecon* or to its likely relationship to Europe in the world of A.D. 2000. To include it simply because of any 'European' tradition it may have seems no more appropriate than to include those other areas of the world where European tradition is the heritage, modified and adapted by contemporary society, like the Americas or Australasia, and perhaps even parts of Africa and Asia where Europe's impact on indigenous non-European society is strong.

We have taught courses on the geography of Europe for many years in the spirit of Ogilvie's *Europe and its Borderlands*, and a revised overview of Europe now seems particularly appropriate to survey and assess the great changes since Ogilvie composed his work a quarter of a century ago laying foundations for changes to come in what remains of this century – the threshold of the third millennium A.D.

R.E.H.M
E.A.S

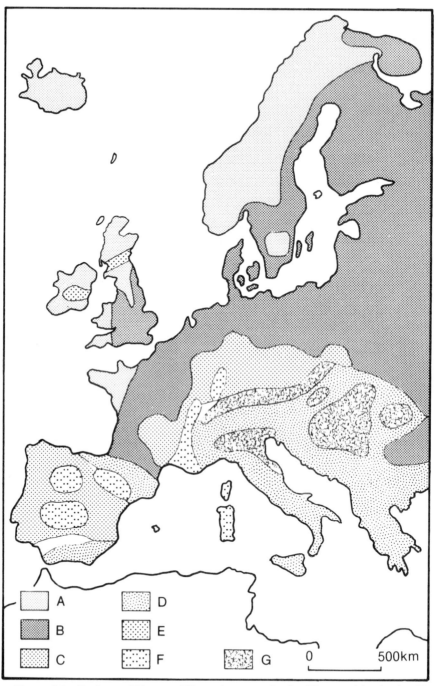

Figure 1.1 Regions of Europe: (A) north-west uplands; (B) central lowlands; (C) central plateaux; (D) Alpine system; (E) lowland regions within A; (F) low relief regions within C; (G) low relief regions within D

1 Introduction: Europe in the World

One of the best ways to understand the significance of Europe's world position is to play with a globe, a tool much neglected nowadays by geographers. We see how Europe – conventionally and traditionally a continent – comprises in reality a collection of peninsulas, islands and epicontinental seas that form the western flank of the immense Eurasian landmass, itself truly the 'world island'. The globe's spherical form shows us clearly how fortunately situated Europe is amid the landmasses that form a mere three-tenths of the surface of our planet. It lies between the Americas and Africa and is firmly attached to Asia; centrally and focally located in the hemisphere that contains the greatest proportion of the earth's land and within this 'land hemisphere' about 95 per cent of human population is encompassed. We also observe that Europe lies at the centre of a circle that encloses the world's richest and most advanced economies. Europe is, however, only third in magnitude of the three great population agglomerations (the others are Sino-Japan and Indo-Pakistan-Bangladesh) that together hold almost two-thirds of mankind.

How do we define Europe?[1] In physical terms, it is a triangular-shaped association of land and continental shelf sea, whose western and southern sides are clearly marked by deep waters, but whose eastern edge, the most truly continental, merges into the vastness of Asia. The long imperfect geographical knowledge of this eastern edge, probably from misconceptions of the true nature of the Caspian Sea and the Volga, gave a belief that water entirely separated Europe from Asia, so warranting without question its description as a continent. When knowledge improved and the idea of an insular Europe was disproved, the status of Europe as a continent was maintained by declaring the Ural 'mountains', a series of puny ranges, as a divide. Many modern scholars would deny Europe's continued status as a continent, whereas others would place its contemporary eastern limits even deeper into Asia than the Ural. There is no rule that forbids a continent from abutting its neighbours – and by what other name could we describe so distinctive a unit as Europe?

Although the physical limits of Europe give rise to debate, as a human geographical concept it is easier to define. For over a millennium it has been a clearly definable cultural area, marked by diverse nuances arising from language and historical experience, but with a common root in ethical values embedded in Christian teachings, however their detailed dogmata may be interpreted. It has been remarkably innovative and has released man from several cultural impasses; but a major factor in making this innovation possible has been European adventure, whether in terms of geographical exploration or in social and political life. Indeed, colonial and imperial adventure has carried Europeans across the seas to other continents, taking with them their culture either to establish their scions as flourishing if modified European societies, as in North America, Australia or New Zealand, or to attract non-Europeans to accept, sometimes with surprising readiness, aspects, artefacts or a veneer of their culture, as in Africa, India and Japan. It remains to be seen whether some of the European scions may not become more innovative and adventurous than Europe itself: they may even reciprocate the dominance exercised by Europeans over the past 250 years over large parts of the earth.

The overseas scions of Europe are readily distinguishable, but the spread to produce new amalgams has complicated the definition of 'cultural' and 'physical' Europe in its easternmost borderlands. Here, there has arisen a state organism basically European in culture that is dimensionally so vast as to dwarf Europe itself – the Soviet Union has modified European concepts in its amalgam with Asia and to fit its own vast scale. The Soviet Union is now oriented as much eastwards into Siberia as it is westwards towards Europe. It is unreal to split it in the belief that 'European' Russia is something different to 'Russia in Asia'. In this work, the Soviet Union is consequently not examined in any depth.

The great age of discovery begun in the fifteenth century by European initiative and inquisitiveness thus lies at the base of many patterns of our contemporary world. Although the political supremacy of Europe, developed in the eighteenth century, was eclipsed by the mid-twentieth century, Europe's influence remains strong, especially if summated through present trends towards multinational and supranational groupings. There seems little reason to believe that Europe's power of innovation and invention are weakening, and in a setting of accelerating technological and scientific progress such an endowment is encouraging for the future. Europe's future prospects appear enhanced by its location at an apex of the massive 'triple continent', where Asia and Africa are showing a new virility for growth.

The dynamism and innovative genius of Europe has been fuelled by the remarkable diversity of physical and cultural environment within such a small area. But many problems have been born of this diversity and a clear awareness and understanding of the great range of physical, ethnic, cultural and economic factors is required for a better appreciation of the growth and development of Europe and of the European idea.

The Continent of Peninsulas

One part of the 'blue planet', when viewed from space, might attract particular interest. There would be a peculiar turbulence of the atmosphere and through the breaks in the cloud-banks would be seen an elaborate pattern of fretted and interdigitated land and sea: Europe. Indeed, whether we take an air satellite photograph or a conventional atlas map, Europe is a continent of peninsulas – many themselves subdivided into more peninsulas, allowing tongues of shallow sea deep penetration into the landmass (table 1.1).

Most striking is the Scandinavian Peninsula, whose western mountain wall makes Norway a truly Atlantic country and much of Sweden a land of the continental interior. In the south, between the deeper basins of the Atlan-

Table 1.1   Composition of the continents

Continent	Total area (million km²)	Peninsulas (million km²)	Islands (million km²)	Proportion of islands and peninsulas to total area (%)	Relation of length of coast to total area
Europe	9.97	2.70	0.75	34.6	1:3.5
Asia	44.18	7.94	2.70	24.0	1:3.2
N. America	24.10	2.04	4.11	25.6	1:4.9
S. America	17.87	0.05	0.15	1.1	1:2.0
Africa	29.82	–	0.62	2.1	1:1.6
Australia	8.90	0.42	1.30	19.3	1:2.0

Source: Wagner, J. 'Physische Geographie', *Harms Erdkunde*, Munich (1976).

tic Ocean and the Mediterranean Sea, is the large square mass of the Iberian Peninsula, with a 'pluvial' Atlantic west and an 'arid' Mediterranean east; while separating the two deep Mediterranean basins is the long 'leg and boot' of the Italian Peninsula, whose volcanoes reflect the structural instability of this young part of the continent; and further east, the broad triangular Balkan Peninsula projects into the Mediterranean basin to terminate in the intricately fretted Greek peninsulas and islands. The main body of Europe projecting westwards from its roots in the Russian Platform may, in itself, be seen as an immense peninsula from which the others project – in keeping with contemporary terminology, we might call it the 'Euro-Peninsula'. Were it not for the relatively recent sinking which formed the Straits of Dover and the southern North Sea, we might still include Britain in the collection of peninsulas.   ' side by side '

In juxtaposition to the peninsulas are the seas. Most impressive is the contrast between the immensely wide and shallow continental shelf in the northwest and the deeper Atlantic and Mediterranean basins in the south. The continental shelf is structurally an integral part of the continent which has been drowned mostly within the past 10 000 years and evidence suggests that it is still continuing to sink: for a period in late Quaternary times, the Baltic was a freshwater lake. The shallow epicontinental seas are important reservoirs of warmth, though the Baltic is relatively shallow and fresh enough to freeze in its more constricted arms of the Gulf of Bothnia and the Gulf of Finland. The considerable tidal variation of the North Sea and its liability to sudden storms aerate its waters remarkably well and it is fed with nutrient by the many rivers that drain to it. But most significant is the penetration of vast quantities of warm water carried from the tropical Gulf Stream by the North Atlantic Drift, whose long fingers reach into the North Sea, along the Norwegian coast and into Arctic waters off northern Norway, even as far as the Kara Sea.

We need only look at the winter problems of Canada's Atlantic coast and at conditions in southern Greenland at similar latitudes to realise what Europe's dilemmas would be if this life-sustaining warm drift were absent.

A major world structural belt – an interaction face in plate tectonics[2] – runs east–west across southern Europe, producing orogenic phenomena of declivities and upfolds, so that in tectonic terms the mountains and seas are closely related. The belt of deep water here comes close to the Pyrenean coast in the tongue of the Atlantic deep that reaches the southern Gulf of Biscay, while three deep basins can be identified in the Mediterranean Sea, separated by mountainous peninsulas or islands. The islands constitute clues to Europe's structural history – the Balearic Islands are a continuation of the Tertiary mountains of southern Spain, but Corsica and Sardinia are mostly remnants of ancient sunken structures, whereas Cyprus comprises deep oceanic deposits. A further deep basin, the Euxine, underlies the Black Sea, while two other deep basins form the major elements of the inland Caspian Sea. A vital fracture in the bow of the Tertiary mountains of the Rif and Sierra Nevada leaves a submerged sill that permits not only navigation into the Mediterranean but also, more important, Atlantic water to stream into the inland sea to replace the heavy loss by evaporation. Nevertheless the Mediterranean is remarkably saline, which increases its ability to retain immense quantities of warmth. An equally fortuitous accident allows water movement between the Mediterranean and the Black Sea, in whose stagnant and badly aerated lower levels petroleum may now be forming. The link across the Manych Depression to the Caspian Sea was broken, probably in Tertiary times, and the latter's salinity is now held constant only through the fortunate occurrence of the shallow natural 'evaporating pan' of the Kara-Bogaz-Gol.

In winter the seas lose their warmth more slowly than the land, whereas in summer they warm up less quickly than the land; consequently, they provide a warming effect in winter and a cooling effect in summer that helps to moderate the land around. Owing to its configuration of peninsulas, islands and epicontinental seas, large areas of Europe enjoy a proximity to water that tempers their climate. It is without a deep continental interior like Asia and consequently without great seasonal extremes of heat and cold. The deeply penetrating seas augmented by numerous inlets and rivers of modest dimension that drain to them provide excellent routes, and in the past it was possible to sail the seas with simple equipment because navigation need seldom be out of sight of land. The penetration of the seas deep into the land also provides numerous short land routes across the major peninsulas, a factor of great historical importance. West of a line from Odessa to Kaliningrad, only a tiny area is little more than 500 km from the coast, but east of this line, once the Volga-Kama confluence is reached, the traveller finds himself at least 1000 km from the nearest coast.

Truly the physiographic character of Europe is without excesses. Of all the continents, it has the lowest average elevation (300 m) compared with Asia (940 m), North America (700 m), Africa (650 m) or South America (580 m). Its highest peak rises to hardly 5000 m, an elevation exceeded by at least 21 Asian peaks, and it is without the extensive dry, mountain-girt basins and depressions typical of Africa and Asia. Apart from the great Russian Platform,

Europe has an attractive arrangement of plains, modest in dimension, set amid uplands and mountains so broken and dissected that numerous useful routes across them can be found, and only the very highest mountains form reasonably serious constraints to travel. They are high enough, however, for their predominant east–west trend to define contrasts between 'north' and 'south', because the east–west grain tends to prevent free northwards penetration of warm sub-tropical and Mediterranean influences, but equally protects the Mediterranean basin against cold and inhospitable northern conditions. Yet nowhere are they high enough to exclude these influences completely. The east–west grain of the younger mountains leaves the coasts of Europe from southern Norway to the Pyrenees open to the deep penetration of mild, moist Atlantic air masses for much of the year, occasionally broken by incursions of dry continental air from the east. Only the mountains along the Scandinavian spine deflect Atlantic air, leaving the Baltic basin much under continental influence.

The Landscapes

Europe may be divided into its component 'regions' or 'landscapes', although there are numerous variations in the way this division may be done, depending much on the criteria used, whether physical features or human associations – through economy, culture, tradition, historical allegiance or political conditions. Even at a low level of resolution the mosaic becomes confusing in the diversity and number of the units. One division into only the most major units is shown in figure 1.1. Many of the clearest defined of the smaller units arise from the innate good sense of peasants for their terrain, classically expressed in the French *pays*.

Interpreted in the widest meaning, three major physiographical regional types occur – the lowlands, the uplands and the mountains. The component regions of these types show an attractive juxtaposition which, though haphazard, has many conveniences for man. Only in the Russian Plain does a traveller have to go more than about 200 km to find himself in another type of region, whereas over most of Europe the distance to be travelled from one major regional type to another is far less. Although none of the regions is of extreme physical type, some undoubtedly have offered barriers to overcome in maintaining trade and contact across them, but even so the great historical routes of trade carried merchants through all the types of major physiographic region. There are few, if any, areas in Europe where the imprint of man is not clearly visible; it is particularly clear in the extensively reclaimed lands, whether from sea or river, moor or marsh; while long farming traditions, both arable and pastoral, have changed vegetation (especially in forest clearance) that in its turn has modified local climate, with repercussions on soil type and even on minor land forms. This imprint is expectedly most noticeable on the more attractive areas for settlement, whereas in many instances the less attractive lands have a long history of the advance and retreat of colonisation.

Until comparatively recent times, the pattern of life within the regions of each major physiographic type was fundamentally similar and differences were in social and economic detail. The life of the peasant in the Alps of the

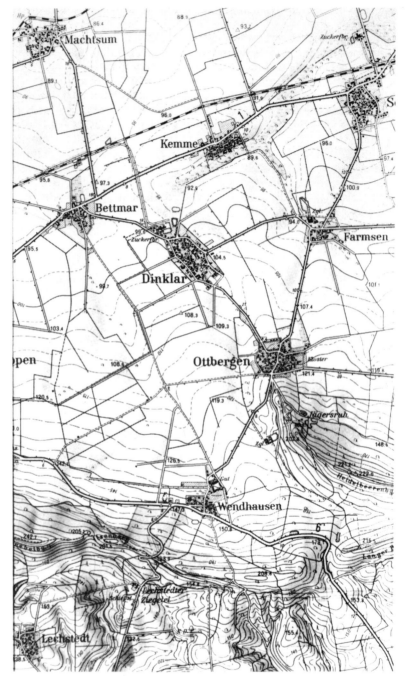

Figure 1.2   Landscape of the central lowlands (B).

Berner Oberland was basically similar to the peasant in the High Tatra, though the latter was poorer, less sophisticated and more oppressed in contrast to the relatively prosperous and well-organised Swiss. The yearly round of the lowland peasants on the *limon* of France would in no way have seemed strange to the Saxon peasants on the *loess* of Central Germany. Nowadays such comparisons are not so straightforward, but the essential truth remains, though it is often concealed, confused and complicated by the ramifications of industry, the dazzle of urban life and the complexity of modern economic and political organisation.

The Lowlands

What are the essential elements of these major physiographic units that together compound Europe? Within its conventional boundaries, well over half the area of Europe lies below 200 m elevation. The larger part of the lowland occurs in the most striking physiographic unit in the continent, the vast triangular area of the Russian Plains. As far west as the constriction in the landmass between Odessa and Kaliningrad, the plain is of immense dimensions, out of keeping with the scale of physiographic units experienced elsewhere in Europe. West of this line the plain narrows appreciably to become the North European Plain, marked on the south by the face of the Carpathians and other mountains and uplands and drowned on the north by the Baltic Sea, extending westwards across Germany into the Low Countries, where it curves round into northern France. We may also think of it extending west into lowland eastern England, for it is in reality shallowly drowned by the late incursion of the southern North Sea. This extensive lowland tract, under several names, may also be considered to reach the Atlantic coast lowlands of France in the west and also into the ancient Fenno-Scandinavian shield lowlands in the north. Elsewhere, lowland occurs amid complexes of mountains and uplands, as in the mid-Danubian Pannonian Plains and those of Moldavia–Wallachia, or in the relatively small central Irish lowland, the Plains of Lombardy in northern Italy and the small coastal plains of Atlantic Iberia.

Generally most attractive to settlement, the lowlands are gently undulating to flat, so that farming is facilitated, but prosperity and fertility are very much dependent on the surface materials and the related soils. We can usually readily distinguish between glaciated and unglaciated lowlands as an important key to their attractiveness to farming. The importance of soil is well illustrated in the North European Plain, where high rural densities and prosperous farming are found on the loessic soils, but densities and prosperity drop sharply on the glacial sands and clays. Good husbandry, manuring and fertilisers, have often improved the soil, but this has depended on the skill and resources to undertake such work; for example, in the Polish plains the soils in former German territories were extensively improved by advanced farming, whereas in the old Russian territories poverty and ignorance of the peasants and disinterest by the landlords did little to improve the soil and farming methods not uncommonly acted to the contrary.

The casual observer may sense a haphazard scatter of settlement in these

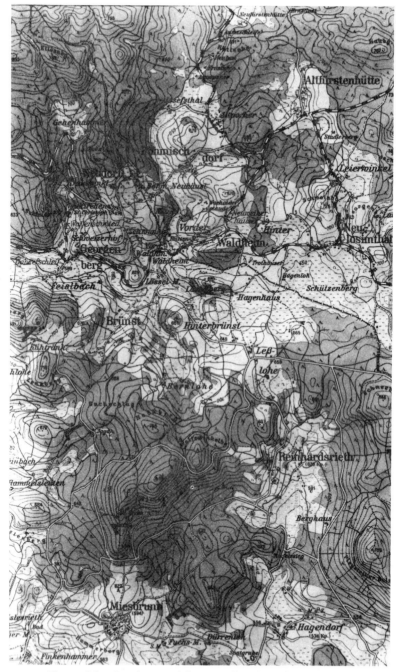

Figure 1.3   Landscape of the central European plateaux (C).

lowlands, but closer examination will show that factors of relief, however gentle it may be, protection and water supply affect the site of each community. In wet glaciated lowlands, settlement is seen to seek drier sandy islands, whereas small mounds – natural and artificial – or dikes enclosing polders are a favoured place for settlement in reclaimed lands. In the great river valleys, villages are also found above the likely flood level, often on gravelly terraces, so well marked in the Plains of Lombardy. Water supply is commonly reflected in the lines of villages along a junction of rocks where an aquifer gives up its water, though artesian water – as in the London and the Paris basins – is also a significant source. In the Russian Plains, of such different dimensions, the marked phenomenon of the high, dry right bank in contrast to the low, meadow bank, flooded in spring by swollen rivers, is a key to settlement, whereas in the open steppe, shelter is sought from bitter winds in the deep erosion gulleys, where strings of villages may extend several kilometres.

Though physiographic conditions may be conducive to similar activities in different regions, in the end the visual landscape may be quite different – the product of different social and political responses. The good arable soils of parts of France and Germany compare readily with Midland and Eastern England, but in the latter social history has resulted in a landscape of hedgerows and compact farms standing amid their fields, in contrast to the open fields of strips, devoid of hedges, with the farms grouped in clustered villages, the visual landscape of much of France and Germany.

Our modern landscape may be so much the product of man's impact on the ecological balance that it is difficult to reconstruct the older environment. For example, in the Hungarian Puszta, part of the mid-Danubian Plains, much debate centres on whether its steppe-like character was created by man's grazing herds from an originally more forested or park-like vegetation.[3] We certainly know that in recent times the Puszta has changed from a pastoral to a grain economy and then swung back to a new importance for livestock. Certainly in any of the major lowlands, climatic factors are important in setting the scene. Even allowing for variations in soil type, in tradition, in market factors, an agricultural map of Britain shows markedly the influence of climatic conditions – notably precipitation, humidity and sunshine – on the livestock and dairy farming of the Cheshire Plain compared with the strongly arable character of the East Midlands. A similar change in pattern can be established across the North European Plain between the dairying of Normandy, the grain and sugar beet in Central Germany or the rye and potatoes of Eastern Poland. The modern landscape may arise from the pitfalls which the success of cultivation has set for man. In Wallachia and Moldavia the growth of sedentary farming from the eighteenth century onwards led to a substantial rise in population and colonisation from the uplands. The growing population became increasingly dependent on the prolific harvests of maize, until by the early twentieth century this was one of the most rurally overpopulated and nutritionally poor areas in Europe, whose falling standards were entangled in a vicious circle of disease and malnutrition. An analogy may be seen with the heavy crop of potatoes that could be raised in the moist maritime conditions of the Irish lowlands, until the disastrous crop failures of the 1840s drove thousands overseas, a migration from which the country is still recovering demographically.

We must recognise that the importance of any region is today not conditioned by its surface features alone: this is particularly true of the lowlands through the coalfields that underlie them in the British Isles and along the northern edge of the Hercynian uplands in Germany and Poland, while salt is widely distributed (for example, in Cheshire and in North-west Germany) and also natural gas and petroleum (for example, in the large natural gas deposits of Groningen in the Netherlands). These buried riches have attracted industry and consequent dense settlement to many parts.

Mountains and Uplands

Europe's mountains are of interesting diversity, though as a generalisation it is possible to say that the older they are structurally, the more modest their elevation. The most ancient structures from early geological times are represented only in the low northern shield scraped by Quaternary ice sheets to leave a landscape of bare rock, forest, marsh and water-filled hollows. The oldest recognisable mountains are comprised of uplifted remnants from the Caledonian orogeny (c. 400 million years ago) that possibly stood originally along the eastern edge of a long-vanished Atlantic landmass. Today they form the mountains of Norway and Scotland, extending into Ireland, where their structures underlie the lowland. The contemporary mountains are only the roots, because the original structures have been eroded to a peneplain, uplifted and eroded again, probably several times over. The last uplift renewed the cycle of erosion, at first possibly under humid tropical conditions, but the landscape we see has been moulded by Quaternary ice sheets and glaciers: permanent ice still remains in Norway on the highest surfaces.

Some 250–300 million years ago, another orogenic phase created broad crescentic sweeps of mountains on an east–west alignment. The story of this Hercynian (including the Armorican and Variscan) orogeny is similar to that of the Caledonian, but today the uplifted fragments of these Hercynian structures, much modified in Tertiary times, are mostly of 500–800 m elevation forming broad tracts of hills and uplands, though they rise to over 1800 m in the Central Massif of France. Most recently, about 50 million years ago, in Tertiary times, another mountain system, again east–west in alignment, appeared across southern Europe to form the Pyrenees, the Alps, the Carpathians, the Dinaric mountains and some ranges in the Balkan Peninsula. It is among these that we find the greatest elevations in Europe – in the Western Alps, seven peaks exceed 4000 m (in the Asiatic part of this orogeny, Everest rises to almost 9000 m).[4]

The Tertiary orogeny is represented by mountains with sharp, pyramidal peaks and a rough and jagged appearance, high enough to carry permanent snow south to 43°N. The Caledonian mountains and Hercynian uplands, in contrast, are landscapes of rounded even summits, with broad open surfaces, remains of ancient erosion surfaces. In the Caledonian mountains, valleys were deepened and straightened by ice and some rounded dome-like peaks eaten into by ice to form corries. Where overdeepened glaciated valleys meet

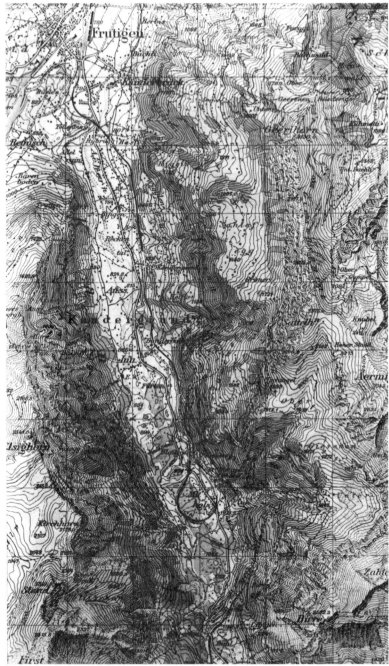

Figure 1.4   Landscape of the Alpine system (D).

the sea we find the Norwegian fiords and the Scottish sea-lochs. Bared by scraping ice, poor acid soils occur only in hollows or the lower valley sides. The Hercynian uplands were generally too low to be glaciated, apart from some of the highest surfaces in the north and west, and mostly too far south to be covered by the great northern ice sheets. They were greatly disrupted by the Tertiary orogeny, whose paroxysms along their flanks had broken and moved them. In Iberia, broad Tertiary peneplains were uplifted into high plains, broken by tilted dislocated blocks; in South Germany, dislocation of the Hercynian basin covered by younger sediments opened the latter to erosion into a classic landscape of scarp and vale; whereas in the Rhenish Uplands we see how rivers like the Mosel and Lahn, on an uplifted erosion surface, have incised their meandering valleys in search of a new base level. Fracturing and dislocation also opened these structures to volcanic activity, so well displayed in the volcanic landscapes of the Auvergne or the Eifel.

containing chalk or lime Although the older structures are mostly crystalline or metamorphic rocks, except where covered by younger sediments, the Tertiary mountains contain much calcareous material. The crystalline and metamorphic rocks let precipitation run from their slopes or remain on the surfaces in peaty or boggy hollows. The calcareous materials allow moisture to seep into them, like other sediments, so that the Tertiary mountains have a generally drier appearance. Where they lie near the Mediterranean, summer drought and low precipitation *emphasise* accentuate the large areas of limestone to make them appear peculiarly arid and inhospitable, as in the Dinaric mountains, peppered by solution hollows and caverns, forming the *Karst*. Unlike the other Tertiary ranges, the Dinaric mountains form an inhospitable obstacle to movement across them, having few transverse valleys and rising steeply from the Adriatic coast. Elsewhere, in the Alps and Carpathians, the fortunate combination of longitudinal and transverse valleys provide important routeways, even through the lower and more rounded eastern ranges, while everywhere convenient passes link the valley systems together.

For man, relief in the mountains and uplands in terms of altitude, aspect and angle of slope is significant. Altitude generally inhibits settlement – on high mountain pastures there is often only a summer occupancy, which still remains vigorous in the more truly peasant economy of the eastern ranges but has declined in the more advanced economies of the western mountains. In the Hercynian uplands, permanent settlement is still found at high altitudes on bare open surfaces, but downhill retreat has already begun. In the northern Caledonian ranges, permanent settlement has already retreated to the lower elevations. Settlement density everywhere decreases with altitude, except under particular historical circumstances in the Balkan mountains. Table 1.2 illustrates altitudes of farming in Europe's mountains. Aspect is also important, with the contrast between the valley side long in shadow and that open to the sun, frequently described for the Alps, Carpathians and other mountains, but it is also applicable in the Hercynian uplands, where land use – for example, the vine on sunny slopes and forest in the shade – is commonly an indicator. Aspect is also significant in terms of precipitation: western faces tend to be moister than eastern slopes in the lee and rainshadow of the summits, as seen in the marked contrast between the moist Atlantic faces of the

Central Massif of France and the dry, sunny eastern slopes; in the Rhenish Uplands, the high western Eifel is more raw and moist compared with the

Table 1.2   Maximum elevation of arable cultivation in upland Europe

Upland region	metres
Østerdal (Norway)	500
Gudbrandsdal and Valdres (Norway)	650
Harz	600
Erzgebirge	1030
Schwarzwald	1120
Pilztal	1160
Urner Alpen-Realp	1540
Berner Alpen-Kandertal	1560
Ötztaler Alpen (North)	2014
Ötztaler Alpen (South)	2083
Findelen (by Zermatt)	2100

Sources: various.

more easterly Westerwald. In the lee of the Central European Uplands are found dry – sometimes unusually drought-prone – basins, such as the Goldene Aue or the Bohemian Polabí. Iberia well demonstrates the contrast in precipitation – the well-watered Atlantic coasts and the backing mountains contrast with the high interior tableland and dry eastern coast. Sometimes, mountains in an otherwise dry setting are high enough to trap high, moist air currents, such as the action of the Sierra Nevada of providing moisture for south-east Iberia.   prevent

Quite modest slopes preclude arable farming and machines can be used only on the gentlest slopes. Where slopes are steep, elaborate terracing, often done laboriously by hand, is needed. In the scarp and vale landscapes of the Paris basin and South Germany, the importance of angle of slope is seen in the arable and orchard of the gently sloping floors of the vales, the vines of the terraced lower slopes beneath the scarps, on whose steeper faces forest clings. In the mountains, the combination of angle of slope and aspect combine as important factors influencing avalanche paths.

Rivers

By comparison with the greatest rivers of the world, Europe's rivers, except for the Volga, Danube and Dnepr, are comparatively short, though their volume in relation to their length is high. The rivers that rise in the areas of high precipitation in the western uplands and mountains generally have a remarkable volume in relation to their length and basin area compared with the rivers rising in the east, where precipitation is generally lower: the Rhine has a flow considerably greater that the northern Dvina, whose drainage basin is twice as large, and the Danube at a maximum flow carries a volume well in excess of the Dnepr and Don combined. Short rivers of considerable volume and marked seasonal regime characterise Scandinavia, whereas long rivers with only short periods of high volume mark Iberia. By 'continental' standards, Britain's rivers are puny, but we may recall that the Thames and streams draining to the Humber were in early post-glacial times still tributaries of an enlarged Rhine–Maas system. The large calcareous mass of the Dinaric mountains produces an unusual pattern of drainage, with few surface but many underground streams. Geological and soil conditions, as well as climate, are important factors affecting run-off and consequently the regime of rivers, which in turn affects their use for man. In the Mediterranean basin, the dry stream beds of summer become raging torrents at times between autumn and spring: in the Italian peninsula, many streams are in fact called 'torrente'. A strong seasonal flow marks the rivers of the Russian Plains, where the long frozen period of winter is broken by the great floods of the spring thaw, followed by low water in summer.

The rivers of Central and Western Europe are less seasonally regular and some have regimes compounded of several elements, such as the Rhine and the Danube – two of Europe's most navigated rivers. Compared with the Russian rivers they are frozen for only short periods, particularly the Rhine. In the upper reaches of alpine rivers, melting mountain snows give a summer high water, though the flow on the Rhine is regulated markedly by the Bodensee (Lake Constance). Low water is generally reached in late summer to early autumn. Downstream, high water also occurs in late winter to early spring, but in summer thunder rain causes local floods, and floods may also arise from surface thaws in winter caused by the passage of warm depressions inland, especially on the Rhine. Considerable year-to-year variation in flow occurs, depending on climatic conditions. The Rhine and other North Sea rivers suffer floods when onshore gales pond back their outflow, though this is also a recurrent mechanism of floods on some rivers draining to the southern Baltic. Both the Rhine and the Danube have regimes complicated by their courses of narrow gorge sections, with steep gradients alternating with long gently graded reaches. The low watersheds between the major rivers of Europe's plains, especially in Western Europe, have allowed canal building to take place relatively easily, so increasing the mobility of river traffic. Plans for canals as links between river basins also exist in Eastern Europe, though little has yet been undertaken, except in the Volga basin.

Climate

The greater part of Europe lies in temperate latitudes, though its most northerly points project beyond the Arctic Circle and its most southerly lands lie below 36°N, and it forms the western peninsular cluster of a large landmass adjacent to a relatively warm ocean. In themselves, these are telling factors in the climatic range to be expected. Perhaps one of the most important factors is the massive drift of warm tropical Atlantic seawater that washes Europe's western coasts, such a marked contrast to the other side of the Atlantic where a cold current sweeping from the Arctic basin extends down into comparable latitudes. Europe is for its latitude anomalously warm and is crossed by a procession of moist, warm air masses drifting eastwards from the Atlantic, slowly to surrender their moisture. The sum total of the several factors which combine to form the European climatic environments has created circumstances in which extremes are rare, climatic restraints are relatively few and as a consequence man has been able without significant difficulty to colonise to much higher latitudes than is commonly found elsewhere in the world.

It is, of course, the relative size and strength of the several different types of air mass that in their seasonal variations give Europe its climate and the day-to-day variation that is weather. The dominant element is the zone of contact between air masses, the front, particularly important along the ocean margins in winter. The variety of weather is also affected by the complex interplay of land and water masses and their different thermal characteristics, because the deep landward penetration of water masses is a significant factor in the avoidance of extremes, which makes Europe so acceptable to man.

The contrast in temperature between land and water is greatest in winter, when the western coasts are washed by warm Atlantic waters. Over Western Europe, warm moist oceanic air mixes with dry and very cold air forced out of the Siberian high pressure system. The interplay of these air masses creates turbulence to give variable wintry conditions. Over the northern Mediterranean, cold dry Siberian air meets warm tropical air, moist from the Atlantic, and the resultant fronts give characteristic winter precipitation. In summer, warm dry tropical continental air from Asia or Africa extends over the Mediterranean to bar the way to moist Atlantic air, leaving the area without precipitation. To the north, warm or cool but dry continental air mixes with moister Atlantic or Arctic air and shallow depressions drift limply eastward to give moisture, much of it from instability thunderstorms.

The gradation of climatic type and the complicating variable of weather make it hard to define limits between 'climatic types', if such really exist in Europe. Nevertheless, abrupt changes over short distances may exist locally, as any traveller in spring or autumn may best experience between the ends of any of the great Alpine tunnels; but everywhere aspect, altitude and local terrain conditions are important.

The western islands and coastlands have notably variable weather from the passage of frequent depressions. Winter is generally mild, except when penetrated by inroads of bitter Siberian air. This is a moist time of year, with rain and snow (often in blizzards). In the British Isles, as in Scandinavia, the western mountains wring much moisture from the passing air, so that in their lee the country is drier but also colder, with the contrast greatest in Scandinavia. Summers are also moderated and moistened by the inland drift of oceanic air, with temperatures mounting inland, especially when under the influence of tropical air and steady anticyclonic conditions, though even in sheltered valleys in the northern mountains, temperatures may rise to over 25°C. In such variable conditions, weather rather than climate dominates and a 'typical year' is hard to define. 'Stimulating' and 'invigorating' are adjectives commonly applied to this belt.

One of the most distinctive climatic patterns in Europe is associated with the Mediterranean basin and its littoral, putting a strong impress on landscape and human activities. Its areal extent inland is limited by the mountains that line that sea's northern shores, as well as the strength of more northerly air masses. In winter a low pressure trough along the northern Mediterranean allows the passage of rain-bearing depressions from the Atlantic, with rain in showers between periods of clear blue sky. The weather is often cool and snow is by no means uncommon in the mountains. Relief features play an important part in channelling air, especially inroads of cold northern air that streams down valleys or spills over mountain rims, giving the *mistral* of the Rhône valley or the *bora* of the Adriatic coast. The usual mental association of the Mediterranean is with summer heat, when hot dry air masses from Asia and Africa hold sway, excluding rain-bearing air, though sea breezes at least temper the heat on the coast. There is a tendency for persistent north winds to develop, weak in the west but strong in the powerful Etesian winds of the east, whose very dryness makes them feel cool. It is the reliability of summer conditions – the dryness and the sunshine – that have made the Mediterranean so recreationally important. Iberia, almost a continent in miniature, has Atlantic properties in its western and northern coasts, whereas its eastern coasts enjoy Mediterranean conditions. On the high plateau of the Meseta, cold sunny winters (with snow on the mountain's ribs) are followed by hot dry summers. The Balkan Peninsula, with high mountains separated by basins, allows cold wintry influences far south, though in summer some of the basins become as hot and dry as the Mediterranean, whose influence also extends into the mountains. Poor summer precipitation and extensive limestone areas intensify the impression of dryness and heat.

Central Europe has a climate transitional from the Atlantic seaboard to the truer continentality of the Russian Plains, but once again aspect and elevation are highly significant local influences. Eastwards, winters become cooler and snow lies longer, whereas summers are warmer and thunderstorms more common. In Hamburg or Rotterdam the climate is essentially maritime, by Berlin or Prague continental influences are beginning to be felt, and by Minsk or Lvov a real continentality is experienced. The most diverse patterns are found in the uplands and in the mountains, whereas sheltered basins enjoy dry conditions and great summer warmth, though cold air drainage in winter makes them sometimes very cold. The classic area of this type is the Pannonian Plain, where conditions are similar to the southern Russian steppe. The mountains in winter are often islands of warmth amid lowlands of cold air – this phenomenon is strongly marked within the solution hollows and basins in the Dinaric mountains, where some of Europe's lowest winter temperatures are recorded. The winter ridge of high pressure over the Central European mountain belt is an important factor in bringing clear, dry air of great stillness, with the higher slopes well above valley mists, providing ideal conditions for winter sports. In summer the frequent incursions of the Azores high pressure system into the mountain belt also brings clear, sunny weather, endorsing the recreational importance of this belt, though instability produces thunder.

Warm summers are also experienced in Scandinavia, cooled on the west by Atlantic waters, which help to warm the Norwegian coast in winter, when the Baltic basin is in the grip of intensely cold conditions sufficient to freeze the Bothnian and Finland gulfs. Precipitation is generally lower in the east, where

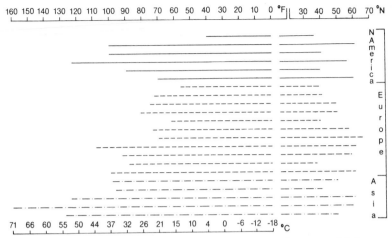

Figure 1.5 Temperature range between January average lowest and July average highest illustrating the moderation of European climates

snow may fall in a light shower for days on end. The long days of summer in such high latitudes allow long solar insolation to balance the short winter days. The continentality of Eastern Europe in the Russian Plains makes seasonal change between winter and summer and vice versa rapid, creating problems for the farmer.

With such conditions of diversity without extremes, all expressed in modest dimensions and readily encompassable by the human mind, it is perhaps not surprising that Europe has proved so attractive and successful as a home for man.

References

1. W. H. Parker, 'Europe: How far?', *Geogr. J., 126* (1960), 278–97.
2. A useful general introduction to plate tectonics is N. Calder, *Restless Earth – A Report on the New Geology*, B.B.C. Publications, London (1972)
3. For a discussion of the origin of the Puszta see N. J. C. Pounds, *Essays on Eastern Europe*, University of Indiana, Bloomington, Indiana (1959).
4. The nature and dating of the orogenic cycles are discussed in A. Holmes, *Principles of Physical Geology*, 2nd edn, Nelson, Edinburgh (1972).

2 The Europeans

The diversity of Europe's peoples and their languages, religions and cultures has contributed greatly to the fascination of Europe and to the emergence of the European culture and ethic which, though frequently modified, has become the dominant influence in many parts of the world. Equally, that same diversity among the Europeans was largely responsible for the nineteenth and twentieth-century emergence of nation states. This in its turn was a major factor in the outbreak of two world wars within a single generation which were to stimulate the post Second World War recognition that increasing unification and interdependence of these diverse elements was essential to the greater well-being of Europe and of the world.

So far as we know, man is a migrant into Europe from one or more of the possible hearths on which the human species evolved. He is most likely to have entered from the south-east, though some consider a possible south-western entry. We cannot date exactly the entry of the first of these human creatures, but it is patently extremely ancient and precursors of *Homo sapiens* seem to have been well established by the earlier glaciation of the middle Pleistocene. These creatures were thus considerably earlier than the short, massive, chinless and brow-ridged *Neanderthaler*, which was possibly a specialised type developed on the utmost limit of the habitable world. We – *Homo sapiens* – appeared in Upper Palaeolithic times, say around 30 000–40 000 years ago, and even at this distant time several distinct physical types were beginning to evolve. Population was numerically small – perhaps a few tens of thousands over all Central Europe and no more than a few hundred in Britain. The early men are distinguished until historical records become available only by their remains as 'cultures', but cultural change may have been forced by the fairly rapid succession of climatic changes – with consequent vegetational and faunistic modifications – in late glacial and early post-glacial times, when the retreat of the ice allowed a slow colonisation further north. By Neolithic times, man was already beginning to have an impact on landscape through his rudimentary cultivation of the natural environment: it is perhaps significant that farming appears to have spread from the south-east, through the Balkans and Danubia into Western Europe. Some influences possibly entered from the south-west and by diffusion around the Mediterranean Basin. Certainly metal-working and pottery-making seem to have spread into Europe from the highly advanced cultures of Mesopotamia via Anatolia and the Balkan–Danubian routeway.[1]

Cultural diffusion was often a product of migration, commonly of a conquering aristocracy but also of large socially balanced groups. Through this process, mixture, assimilation or extinction marked the compounding of physical types that slowly produced the anthropological pattern of Europe, though some change was by natural selection towards the most economical physical form (for example, the increasing frequency of roundheadedness). Through this process, somewhere around the middle of the second millennium B.C., Indo-European languages began to spread, supplanting earlier tongues, though little is understood of the process of and the reasons for language diffusion.

'Race' is essentially a physical anthropological concept: the attributes most commonly used are pigmentation, head shape, stature and facial characteristics, whereas some modern studies use blood groups and even senses such as taste and smell. Unfortunately, for past populations there are usually too few remains to allow large samples, and skeletons tell us little of the appearance of the people's hair or skin colour; for some periods (for example the late Bronze Age) cremation has denied us even skeletons. The whole question of race has been much complicated and confused by emotional and subjective evaluations, with common confusion between physical anthropology and ethnology (or the modern social anthropology). Early work on physical race was concerned with individuals rather than populations, to which more recent interest has shifted. In any population there is only exceptionally any physical unity and usually members of several physical types are represented – though the proportions of the several types vary from one group to another.[2]

Simple and commonly used classifications of the major racial types in Europe have been devised which divide the Europids – 'White Caucasians' – into Nordic, Dinaric–Alpine and Mediterranean types, though it has been suggested that these are, in reality, variations of a single group and result from adjustment to environment and towards an ideal form. From these major types, some smaller groups have been distinguished by anthropologists, including remnants of older stocks, as well as sub-types (for example the East Baltic type, a Nordic–Dinaric Alpine cross). The most marked feature of the Europeans is their low level of pigmentation, decreasing northwards, though in fact 'whites' are found autochthanously beyond Europe, with an apparently much wider distribution in the distant past. The low level of pigmentation is thought to have taken place in the early phases of human colonisation northwards: to reach the present level, anywhere between 5000 and 100 000 generations may have been needed, though it could have been helped by social favour shown to light-skinned and light-haired individuals, since concepts of beauty and ugliness as selection factors are important in the evolution of a race.

Northern Europe, notably Scandinavia, but also the south Baltic littoral, has a high frequency of blond, blue-eyed, light-skinned people, generally with long heads and tall stature. These are members of the Nordic group. In the south-east Baltic littoral, shorter stature with broader heads and often flattish faces and high cheek-bones forms a sub-type. Migration, notably by the Germanic tribes in late Roman times, carried these characteristics widely into more southerly areas, while the Scandinavian Dark Age seaborne migrations carried them along the Atlantic coastlands. By far the most widespread type, especially in Central, Eastern and much of Western Europe, has medium stature, is brunette but is nevertheless fairly light skinned (with brown or light-coloured eyes) and medium to broad head. This comprises the Dinaric–Alpine group, widely disseminated through migration. Southern Europe contains a high proportion of generally broad-headed people of medium to short stature, whose hair, eyes and skin are generally dark to swarthy. It is sometimes difficult to distinguish these people from others living around the Mediterranean outside Europe. Some of these southern traits can be found among the populations along the Atlantic seaboard, even into southern Scandinavia, most likely from early prehistoric migration. The likely survival of early physical types in upland and inaccessible coastal areas in the remoter western areas or in the more inaccessible mountain areas of Central and South-eastern Europe is partly supported by studies of these populations. Some Asiatic traits can be found among peoples known to have entered Europe from Southern Russia and the Volga basin.

Evidence suggests that the various 'elements' of physical race are independent and not interdependent, so that in Europeans there are multifarious combinations of form and colour centred on three basic characteristics inherited from the past, rather than arising from mixtures of three ancient pure races long since mongrelised. For example, there is the continuation of light pigmentation, accentuated in the north, most notable among older, larger bodied and larger headed stocks derived from neo- and mesolithic peoples, whereas roundheadedness is widely prevalent, partly through genetic change, partly through migration, especially in a broad middle zone trailing eastwards into Asia. It would be misleading to consider the pattern and forms so established as immutable: 'racial' change is continuing, notably in our own time by the introduction of non-European stocks into several parts of the continent.

Blood group study suggests that the frequencies show, whatever other characteristics are found, that the earliest peopling of Europe was by successive waves from the east and south-east. In this respect, the Basques are interesting, since they speak a non-Indo-European language of exceptionally ancient origin and, as a population, display low A, very low B and high O blood group frequencies, patterns discernible in some other isolated areas of the Atlantic fringe. Local variations in blood group frequencies that suggest migratory influence have been found, as in contrasts between coastal and inland populations in Northern Europe, but much more work on blood groups and their frequencies is needed before certain conclusions can be made.

'Race' in the physical sense has played little or no part in the emergence of a sense of community among peoples in Europe. Some students in the nineteenth century tried to claim superiority for certain physical types (notably the Nordics or the 'Aryans'), but it was not until the Nazi racialist philosophies of the 1930s that such thinking received any official credence. Even widespread antisemitism should be seen more in a cultural–ethnic light than in a real 'racialist' sense. Racialist feeling has, however, generated tensions in some areas owing to considerable recent immigration of non-Europeans (for example, people from the Indian sub-continent and Africa or the West Indies

into the United Kingdom, Indonesians and other Asians into the Netherlands).

The emergence of a sense of community and consequently national identity
has been centred overwhelmingly on ethnic factors – combinations of language, religion, social usage and other facets – besides historical experience
and symbolism. From the sixteenth century, the spread of printing and the decline of the Latin scribes gave a new importance to vernacular languages
around which national identity began increasingly to crystallise, encouraged
in the eighteenth and nineteenth century by the study of folklore and folk
song.

Since early historical times the distribution pattern of language, like the
languages themselves, has changed: indeed, the spatial patterns and structural forms of languages are dynamic and constantly modify in response to
social, economic and political conditions. It is thought that by Iron Age times
the main language groups had evolved and if one common Indo-European
language had ever existed, it had already divided into two main groups–the
satem and *centum* (based on the word for *hundred*) types reflecting an easterly
(mostly Slav) family and a westerly (Germanic–Romance) family.[3] Lithuanian is considered to remain closest to the original Indo-European form. A few
contemporary languages in Europe – Basque, Magyar, Finnish, Karelian,
Estonian and some languages in European Russia – do not belong to the
Indo-European group, being immigrants of various dates. In tracing the
spatial shifts in language, useful evidence is given by place names. For
example, Celtic place names are widely found in Central Europe in areas
from which Celtic speech has long since disappeared, or the Slav place
names in the North German Plain reflect an early mediaeval pattern of
settlement. River names are commonly of extremely ancient lineage. Change
has also been slow: the division of Celtic into two major forms seems
to have taken about a millennium to mature, whereas little change over
several centuries is recorded in the geographical borderline between, say,
French and German speech in Lorraine, or French and Breton speech in
Armorica. On the other hand, rapid shifts in the eighteenth to twentieth centuries have taken place in the distribution of German and Slav speech in East
Central Europe. Language is open to strong social and political pressures: the
emergence of modern English from Anglo-Saxon and Norman French is a
useful example, while the spread of German in East Central Europe under
Prussian and Habsburg hegemony made it a major *lingua franca*, whereas
French as a diplomatic language was intimately linked to France's political
fortunes.

European languages have been carried widely beyond the continent by
migration. English has been taken to North America, where it has become the
predominant language; to Australasia; and as a *lingua franca* through the
former British colonial territories of Africa (including its establishment
alongside Dutch Afrikaans in South Africa); whereas it is also used in Asia as
the major vehicle of communication in the polyglot Indian sub-continent.
Since the Second World War, English has become virtually the first foreign
language throughout the world. Other examples are the spread of Scandinavian dialects by Viking colonists to Faroe, Iceland and Greenland (later to die

Figure 2.1 Languages of Europe

out), while Scandinavian influences are found in many dialects in the British Isles, though French became accepted by the Norse in Normandy. French has been spread by settlement and colonial ventures in Canada, in the West Indies and widely in Africa and Indo-China, and parallel stories for the spread of Portuguese, Spanish and Dutch can be told. The collapse of political power or social status, as much as extinction or slaughter, have been factors in the decline of language: the once widespread Celtic languages are now fighting a vigorous rearguard action for their existence in island and mountain retreats on the outermost Atlantic fringe of Europe. Some, like Cornish, have been almost lost already. Illyrian has likewise disappeared with little sure trace, just as Borussian (Old Prussian) has done in the north, and we know little of Pictish.

We can group the languages in major families: Germanic, Romance and Slav being the most important; of lesser significance are the Celtic languages and even the 'outsiders' (Magyar, Finnish, Karelian, etc). The Germanic languages seem to have evolved around the western Baltic and were spread from late Roman times by tribes migrating south-westwards. A massive spread of these tribes in the fifth and sixth centuries A.D. left some memorials, like the Germanic place name elements in Iberia and in Italy; Angles, Saxons and Jutes carried Germanic languages into the British Isles, later augmented by Viking influence. The northern branch came to form the Scandinavian languages, Norwegian, Danish and Swedish (which also spread to a tenuous hold on the southern Baltic shore and a more lasting but now diminishing presence in the Finnish coastlands), whereas the more isolated Icelandic and Faroese evolved slowly from the language of early Norse colonists. The modern mainland Scandinavian languages are, with some effort, mutually intelligible.

German became the main body of the family in Central Europe, divisible into High, Middle and Low German dialects. Low German also emerged into separate languages, Dutch and closely related Flemish, while Frisian shows close affinities to English. The High German or Alemannic dialects colour everyday speech in South Germany and Switzerland, and they also spread widely through the Austrian dialects used in the Habsburg Empire. Modern formal German is based primarily on northern usage (notably the written language) and pronunciation. German dialects remain everyday speech in Alsace and Lorraine, Luxemburg and in Eupen-Malmedy in Belgium.

After 1919, and still more rapidly after 1945, territorial and ethnic change reduced the German speech area in East Central Europe and the Danubian countries through the dissolution of German colonies existing mostly since mediaeval times. Some islands still, however, remain – notably in the Transylvanian Saxon settlements of Rumania and in Polish Upper Silesia. The large German minority on the Volga near Saratov in the Soviet Union was dispersed to Siberia during the Second World War. Rhenish Jews carried German mixed with Hebrew into East Central Europe in mediaeval times as an important commercial and cultural tongue known as *Yiddish*. German nevertheless remains a major commercial and scientific tongue used widely as a *lingua franca* in East Central Europe (notably Hungary, Czechoslovakia and Poland).

Of the Germanic family, English is the real success story. An amalgam of Anglo-Saxon (a Low German-type language) and Norman French, with Scandinavian and Celtic traces, the modern formal and written language derives from the dialects and usage of the English East Midlands, but many might claim a separate identity for the English used in Scotland. Since the sixteenth century it has undergone a world spread, becoming the dominant language in North America and the national language of Australasia. It is a major *lingua franca* in Asia and Africa as well as in Europe and is used as a main scientific and commercial language the world over, some fields working almost entirely in English (for example, airline operators and shipping associations).

The second important group comprises the Romance or Italic languages, diffused from a core in the Italian peninsula throughout the Roman Empire. Roman civilisation introduced Latin into all parts of the empire and it was used by romanised indigenous peoples, but only in Iberia, the Italian peninsula and in Gaul did it effectively become the basis of language in later times. Classical Latin, a formalised written language, diverged from the racy and dynamic spoken form at an early stage and survived as a language of clerks and scribes. Once the standardising influence of the empire disappeared the regional taint of Latin by local words and usages laid the foundation for separate languages to appear, so that by the time the clouds of the Dark Ages cleared, these regional forms had become separate and mutually intelligible languages. Italian, which has remained closest to Latin, is first recognisable from the latter tenth century. Modern Italian is based on the Florentine dialect, though the finest spoken Italian is said to be *la lingua romana nella bocca toscana* (the Roman tongue in the Tuscan mouth). Several dialects used in the Alps preserve older forms (Ladinian, Friulian and Romansch – the latter being recognised as a national language by the Swiss in 1937). Spanish first appears in recognisable form in eleventh century writings and in all four languages of Romance origin can be discerned in Iberia: Catalan Spanish, now tending to be overwhelmed by Castillian Spanish, Portuguese and related Galician (also under pressure from Castillian Spanish). Portuguese is much 'contaminated' by non-European words, while some words of Arabic and Moorish origin remain in Spanish, a relic of the Moorish occupation.

French, in several early dialects, had spread over much of Gaul by Dark Age times, but it was much tainted by Germanic influence from the Frankish tribes that flooded in as Roman power declined and from which its name was derived. The fourteenth and fifteenth centuries saw the rise to supremacy of the northern dialect of the Île de France over the southern speech, the predecessor of modern Provençal. French became not only the official language throughout the large colonial empire, notably in Africa and Indo-China, but also remained the speech of the French *habitants* of Canada's Quebec province after the British takeover. Most significant, in the eighteenth and nineteenth centuries, French rose to be a world *lingua franca* in diplomacy and an important medium in polite society in Europe. In contrast, Italian hardly spread beyond the limits of the early twentieth century Italian Empire in Africa, and even its use on the Dinaric–Adriatic littoral and many eastern Mediterranean islands has declined. Spanish and Portuguese were carried far and wide by navigators and empire-builders – South America came almost entirely under their influence (Latin America), whereas Portuguese also became widespread in large colonial territories in Southern Africa. They remain important commercial languages.

The third major group is formed by the Slav languages, among which Great Russian has come to occupy a significant world position through the rise of the Soviet Union. Early in our era the Slav people spread from their coreland somewhere between the Carpathian footslope and the Pripyat marshes, moving westwards and southwards in the wake of the vacuum left by tribes plundering into the decaying Roman Empire. In the process three groups developed from what was originally probably a closely related and mutually intelligible proto-Slav tongue. The Western group comprises Polish (and its related Kashubian and Masurian), the mutually intelligible Czech and Slovak (a very late divergence) and Sorbian. The Eastern group comprises Great Russian, Little Russian (or Ukrainian) and Byelorussian. The complex Southern group basically comprises Serbo-Croat (Serb written in Cyrillic, Croat in Latin letters, but otherwise mutually intelligible), Slovenian and Macedonian. The latter is closely related to Bulgarian, a language with its own distinctive features and spoken by a people who adopted Slav language and culture after entering the Balkans.

A strange hybrid between Slav and Romance languages is Rumanian and Vlakh (a tongue used by scattered mountain shepherds in the Balkans) derived from Roman provincials who seem to have adapted their speech to a basically Romance grammatical structure but a markedly Slav vocabulary. In modern times many Rumanian technical and related terms have been adopted from the Romance languages and the ties with them stressed.

The once widespread Celtic languages are now found only in remote and isolated areas: Breton in the French Armorican Peninsula, Welsh in the Cambrian mountains, Gaelic in the western highlands and islands of Scotland, and attempts to create a separate identity for the Republic of Ireland have brought lively efforts to revive the Irish language. Cornish and Manx are being exhumed. Migration to urban industrial employment means that some of the main groups of native speakers are now paradoxically in the big cities (for example, Gaelic speakers in Glasgow). Modern Greek displays comparative stability in location, though it has recently lost ground to Turkish under war conditions in Cyprus, and it depends on the Orthodox Church to keep it alive in many scattered communities. It is as far from Classical Greek as modern Italian is from Latin. Possibly its separate alphabet has helped its survival. The south-east Baltic provides an environment in which ancient languages – Lithuanian and Latvian (Lettish) – survive, despite encroachment by German and Russian. Albanian is a survivor of the Illyrian–Thracian speech, with several distinct dialects. It is spoken in Albania and in wide adjacent Yugoslav territories, as well as in pockets in Greece and even southern Italy and Sicily. Gypsies brought Romany, an Indo-European language, into the Balkans from India about the fourteenth century, though its importance (like Yiddish) has declined strikingly since the Nazi extermination policies in the Second World War.

The non-Indo-European languages in Europe are represented most strongly by members of the Ural-Altaic (or Finno-Ugrian) group. In the north-east, as a result of migration under pressure from more virile groups, Finnish, Karelian and Estonian are found. When Ingrian was widely spoken before the invasion of Russian, the whole head of the Gulf of Finland spoke these tongues. Many speakers were absorbed by other peoples, notably the Great Russians, though small islands (Veps and Vods) still exist in the Valday Hills and some Russian Arctic peoples use these languages. The Lapps and related tribes are still distinguished by their way of life as much as by their language. The Magyars migrated into the Pannonian Plains in the mid-ninth century and though fundamental changes in their way of life have taken place, they retained their language. Magyar was spread in the nineteenth century by a rigorous policy of forced assimilation and is spoken today outside Hungary in adjacent areas lost to Slav states and to Rumania after the First World War, while a distinctive Magyar social and cultural group remains in the Szeklers of Transylvania.

The Basques are one of the most enigmatic groups – their language is a mystery as it bears virtually no relationship to any other known tongue and is regarded as an exceptionally ancient survivor of the pre-Indo-European phase. It has been a vital key to maintenance of a closely knit and cohesive society and separatist movement.

Language has been a major criterion in defining national ethnic groups; consequently, it has become a significant parameter in political geography, so that there has been considerable attention devoted to statistical measurement in censuses, and even policies to force the use of a particular language have been launched in some ethnically diverse states. Political relationships have often forced the choice of foreign languages studied in schools. In some instances, attempts have been made to raise dialects to the status of languages. For example, attempts were made between the world wars to raise western dialects of Ukrainian to the status of a separate language – Ruthenian. Linguistic affinity has also been disputed, as in the Bulgarian–Serb debate over Macedonian. Language has been important in many separatist movements, as among the Basques, while politicians sought to boost the Irish language in the Irish Free State in the national iconography.

A most significant element in discerning ethnic groups has been religious affiliation. Christianity spread from the Levant into Europe and Africa and even quite far afield into western Asia, but in the Islamic explosion, apart from pockets of Nestorians and Coptic Christians, it was largely eliminated outside Europe until the beginnings of the age of colonialism in the sixteenth century provided a new opportunity for missionary work. In the Dark Ages and mediaeval times, the great monasteries not only kept alive learning but also became focal points of economic activity, especially in clearing and colonising land. Spreading from the Italian peninsula through the declining Roman Empire, the first dissemination slowed as more inaccessible regions were reached, with the last conversions in Lithuania and among the Finnic peoples of the north.

The present denominational pattern is represented in figure 2.2. The Roman Catholic church, still one of the most widespread, emerged in the fourth century under the Bishop of Rome and spread quickly through the

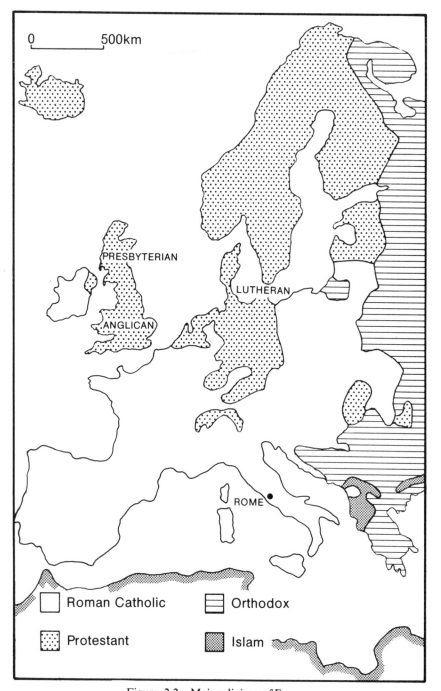

Figure 2.2 Main religions of Europe

weakening Roman Empire. Missionaries from Britain, Ireland and France converted the pagan German tribes between the seventh and ninth centuries and Germans carried the faith to Scandinavia and to the Slavs (Poland converted in the late tenth century), whereas the conversion of the Magyars was completed by 1100. Swedes had carried the gospel to the Finns in the twelfth century. The last conversions were made of the Lithuanians about 1386. For a time, Ireland and the western seaboard of Britain became isolated from the mainstream of Christian development, and during this isolation there emerged the distinctive Celtic church that has left its imprint in the landscape. The use of the Latin alphabet has been an important inheritance from the Roman church.

The Byzantine church had been successful in christianising many Slav immigrants into the Balkans and in carrying the Christian faith to the Russian tribes, especially through Vladimir of Kiev (989); but it had very little success north of the Danube and remained primarily within the contracting boundaries of the Eastern Roman or Byzantine Empire. Expeditions westwards by its missionaries generally failed in the face of a more aggressive Church of Rome. The religious schism of 1054 had, however, very much divided Europe between the influence of Rome and Byzantium, but the rise of Islam and the Turks lost for the latter its successes in Asia Minor. The Byzantine church, particularly after the fall of Byzantium to the Turks in 1453, lacked the strong central organisation of the Church of Rome and divided into regional and ultimately national churches under their several patriarchs and exarchs, with strong Greek and Russian influence. Some Orthodox Christians retained their rites but fell under control of the Roman church, notably through Polish influence in the West Ukraine and northern Rumania. This so-called Uniate church was reunited to the Orthodox church proper in 1945. The Orthodox churches displayed little evangelising and missionary fervour and few conversions were achieved, especially among the non-Christian people of the Russian Empire. The Byzantine influence still markedly coincides with the use of the Cyrillic alphabet devised by one of the church's early missionaries, St Cyril.

The religious upheaval of the sixteenth and seventeenth centuries left Protestant dogmata well established in Central, Western and Northern Europe, though without the rigorous central authority exercised by the Roman Catholic church, the Protestants became divided into many sects. The pattern of distribution between Roman Catholic and Protestant areas has remained remarkably constant from the mid-seventeenth until the mid-twentieth century, though some local shifts in distribution have taken place with migrations generated by industrialisation. In 1945, the largely Protestant population in German territory east of the Oder–Neisse rivers was expelled and replaced by Roman Catholic Poles. Many people have seen the Protestant and particularly Nonconformist philosophy of diligence, thrift and hard work as a significant factor in encouraging industrialisation, and being less conservative than Roman Catholicism it encouraged innovation and social change.

The incursion of Islam left no trace in the religious pattern of Iberia, but Turkish conquests in the Balkans resulted in conversion of some indigenous people to Islam as well as the settlement of Moslem Turks in several districts.

Moslems are still found in Albania and adjacent districts of Yugoslavia, as well as in Bosnia, but many Moslems in Greece and Bulgaria were returned to Turkey in 1920 and after 1945, though the small groups of Slav *Pomak* Bulgarians remain true to Islam. It should be recalled that Moslems are found as far north as Kazan on the Volga. Migration of Turks and Asians as workers into Western Europe in the late twentieth century has established many Moslem communities in large and medium-sized towns. In 1939 over half the world's Jews lived in Europe, where they had arrived after the dispersal from Palestine beginning in Roman times, spreading widely into Central, Western and parts of Eastern Europe. Groups from the Rhineland drifted into Poland and Russia, as persecution in mediaeval times diffused the communities, as also for example did the movement from Iberia to parts of the Byzantine Empire; and in 1804 the Russians defined the *Pale*, so restricting the area in which Jews might live. Nazi racialist policy in the 1930s began an unprecedented persecution, in which large numbers perished or left Europe, many settling in what has become Israel.

Other important ethnic indicators are the arts and music, though they have never been effectively mapped. The musical tradition is particularly strong, especially in Central, South-eastern and Southern Europe, while a distinctive element of Gypsy culture is its music. In both Iberia and the Balkans, non-European influences can be traced in folk music. Folk art has also been important: even into the late nineteenth century folk costume was commonly worn, but nowadays it is mostly seen only on high days and holidays. Many peasant costumes, as for example in the Balkans, can be traced back to prehistoric origins, and they have also had an influence on military uniforms (for example, the square-topped cap of Polish origin seen in Polish uniforms, the German *Uhlan* cavalry and even British lancer regiments, whereas the Russian soldier's shirt-like *Litevka* is a peasant garment). Folklore has been an important element in literature, though often carried into other cultures (many popular fairy stories in Britain are of Central European origin). There remains a fruitful field for geographical study in social usage and taboo.

Major ethnic groups are important factors in people's minds when defining regions as are the often related, but not necessarily coterminous, language distributions. In the north there is the Nordic or Scandinavian province – a collection of related languages (except Finnish), a uniformity of Protestant denominations, and a strong cultural affinity in art and music, reaching back to the Viking age and even to pre-Christian myths. A second major province is a vaguely defined Central Europe, a largely Germanic province, but with mixed religious patterns, yet nevertheless in art and music and folk usage fairly clearly related. Through the influence of Celtic culture and the strong impact of the Normans, the British Isles form another province where distinct art forms and musical traditions can be seen, though a fundamental religious cleft exists and there is a dichotomy between the 'Anglo-Saxon' and 'Celtic' spheres now being accentuated by a synthetic nationalism.

Romance Europe is essentially Roman Catholic, its culture and its life being heavily stamped by this denominational mark. There is a strong unity of language without mutual intelligibility. Music and art are significantly related, though in this field the French are perhaps the most divergent. Nevertheless, within this province there is a marked difference between its eastern and western spheres, though influences from the Mediterranean beyond Europe can be traced in both.

Within Slav Eastern and South-eastern Europe there is great diversity and the province contains certain non-Slav elements, while some observers would put some Slavs (for example the Czechs) into Central Europe, for among them until recently German influence has been extremely strong. Russia, through its long periods of self-chosen isolation, is most distinctive and shows marked Asiatic influences; even so, many Slavs regard 'Mother Russia' as a spiritual and cultural home. The Southern Slavs (along with the Magyars and Rumanians as well as the Albanians and the Greeks) form a distinct group, where a division might even be made between a Danubian and Balkan sphere (with persistent Turkish influence in the latter). Music has played a very significant part in Slav traditions, though there are certainly marked regional variations, and folk art is also readily distinguished. Religion has divided this ethnic province between the Roman Catholic and Byzantine churches, with the important cultural barrier of two alphabets.

References

1. Discussion of the origins and diffusion of early man is found in C. S. Coon, *The Origins of Race*, Cape, New York (1962).
2. The concept and nature of race are discussed in J. R. Baker, *Race*, Oxford University Press, Oxford (1975); J. S. Huxley and A. C. Haddon, *We Europeans*, Cape, London (1935); and J. Geipel, *The Europeans: An Ethnohistorical Survey*, Longmans, London (1969).
3. The development and classification of European languages can be usefully examined in F. Bodmer, *The Loom of Language*, Allen and Unwin, London (1943).

3 The Political Quilt

No understanding of Europe can be complete without some knowledge of how the pattern of the political quilt of nations and states has emerged – part of a continuum of shifting boundaries and allegiances.[1] Some form of political–territorial organisation doubtless existed from the earliest times and even some contemporary ideas of state and nationhood are descended from the small and mostly city-states of Classical Greece, whose activity in establishing colonies around the Mediterranean basin and the Black Sea shores helped to spread urban culture. The great but ephemeral empire of Alexander of Macedon expressed an early form of political–territorial organisation: the large personal empire. It was an early European political venture beyond its own boundaries.

One of the most impressive political–geographical organisations was the Roman Empire, whose imprint on social and cultural life remains into modern times, and there is still a marked difference in life and the landscape between areas that were inside or outside this remarkable organisation. It spread through superior organisation of its armies and of its territory, with an elaborate web of towns and roads the like of which had never before been seen. The 'romanisation' of its subject peoples left its mark in several areas on language, a major element in the later emergence of national identity. Its towns and roads, though much neglected after its demise, were a framework around which several states in early mediaeval times were woven. The Romans carried their 'European' ideas and usages into North Africa and into western Asia Minor, creating the greatest unity around nearly the whole Mediterranean basin – the *Mare Internum* – ever achieved. In its decline through internal weakness, its landscape was so attractive that many peoples beyond the Roman border flooded in to raid its territories or even to settle permanently, again influencing the pattern of peoples and states to emerge in mediaeval times. The final split, in an attempt to save its traditions, has left a cultural divide in Europe to this day – the Latin alphabet and Roman Christianity mark the heritage of the Western Empire, just as the Cyrillic alphabet and Orthodox or Byzantine Christianity mark the Eastern Empire.

In the decline of the Roman Empire, Europe was seized by a great restlessness. One of the most powerful movements was the incursion of Germanic tribes that set up petty states in the Roman image within the decaying carcass. In the last stages, there appeared from the east the first Slav tribes that spilled into the Eastern Empire, seeking new lands in the Balkan Peninsula, which also attracted the Turkic Bulgars from the Volga basin, who were eventually to accept Slav language and customs. There were also intrusions of fierce

Asiatic nomads, among whom were notably the Huns and Avars. Long after the Western Empire had disintegrated into ill-defined petty states, ruled mostly by Germanic aristocracy and punctuated only by brief personal empires of size and transient stability – like that of Charlemagne – a new invasion broke, the incursion of Islamic Moors into Iberia (eighth century A.D.); while for almost two centuries until the tenth century, Western Europe's coastlands and river basins attracted the Vikings, who ultimately settled some of the lowlands, coastal lands and islands. The late ninth century marked the inflow across the Carpathians into middle Danubia of the Magyars, to become permanent residents of Europe. In such turmoil the pattern of political allegiances changed quickly and everyday life was gravely disrupted, while learning retreated into the monasteries, but it was not surprisingly a critical period in the formation of the peoples of Europe.

The Rise of Mediaeval Europe

By the early eleventh century a period of relative stability had dawned and the political map (figure 3.1) had begun to crystallise. For a brief period in the ninth century, the Frankish Empire, under Charlemagne, had united the emergent French and German peoples. The ensuing symbiosis of romanised Gallic population and Germanic aristocracy was marked by the French Kingdom, though the real unity of France had yet to be realised through incorporation of Burgundy and of Normandy, with its absorbed Norse aristocracy. The German peoples, still with strong local allegiances, had failed, however, to find the same political identity as the French and had been swept into the vague but brilliant concept of the Holy Roman Empire of the German Nation, charged by the Roman church to carry on the Roman tradition. For the greater part of its thousand-year existence the empire was in essence to be an epilogue of grandeur and to bedevil the search for and realisation of German national unity. Italy, within the Holy Roman Empire, was a collection of petty states graced by the name of 'kingdom', whereas the south had fallen to Viking rulers and Sicily for a time was held in the name of Islam, which had also spread to Corsica and Sardinia. Iberia still lay largely in Moorish hands, though an emergent Spanish identity in the north struggled to force back Islam. The Scandinavian kingdoms, though fluid in their boundaries, had already emerged as Norway, Sweden and Denmark. For a brief time under Canute (died 1035), a vast Danish empire had held much of England, parts of southern Sweden and large tracts of Norway. The Anglo-Saxon kingdoms of England had been united in the struggle against the Danes under the leadership of Wessex, but a major event in the political geography of Europe was their conquest by the Normans in 1066. Scotland, with strong Norse influence, and Ireland (like Wales a stronghold of Celtic culture) were yet long to remain independent.

The eastwards spread of the Germans, forced by their containment on the west, from the eighth century A.D., had pressed back the most forward Slav groups and the thousand-year struggle between the Germans and the Slavs in Central Europe had begun. A Polish dukedom had been established by the year 1000 and enjoyed brief glory, but the Bohemian dukedom from

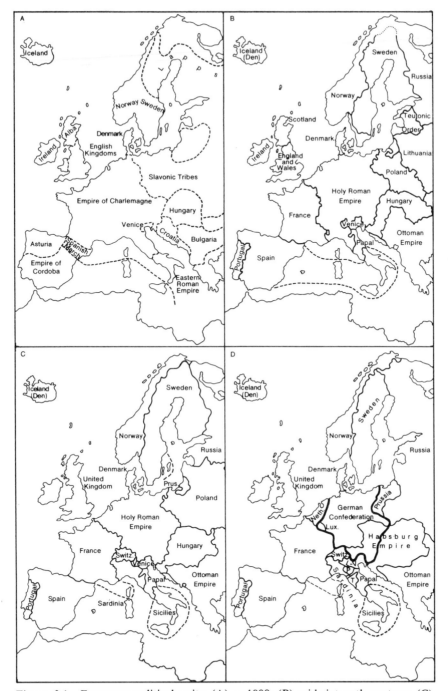

Figure 3.1   European political units: (A) *c.* 1000; (B) mid-sixteenth century; (C) 1793; (D) 1815

inception was strongly influenced by the Germans and lay within the Holy Roman Empire. The immigrant Magyars had quickly become sedentary and the Hungarian kingdom was already in being by A.D. 1000. In South-eastern Europe the rotting remains of the Byzantine Empire had been slowly devoured by the expanding settlement area of immigrant Slavs, while in the mid-tenth century a brief Bulgarian Empire blossomed. The southern Russian steppe was held by fierce Asiatic nomads, and the eastern Baltic littoral and Finland were still heathen tribal lands. Byzantine Christianity had, however, taken hold in Russia through Kiev, where a brief imperium had been built by Vladimir in the tenth century, to disintegrate later into petty Russian principalities.

The Church dominated the life and times: the importance of its place is seen in the remarkable organisational achievement centred on the Holy Roman Empire of the Crusades (1098–1291) as an armed excursion outside Europe to recapture from Islam the 'Holy Places'. When this faltered, attention was turned to christianising within Europe, notably in the Baltic basin, where the German Orders of Knights have left a still visible impact on the landscape, or through the final reconquest of Iberia from Islam. The brief fierce incursion of the Mongol hordes in the thirteenth century (that collapsed because of its own leadership problems) was, however, nothing compared with the storm to follow for three hundred years in South-eastern Europe from the invasions of the Ottoman Turks. Their rapaciousness, disinterest in trade and economic development and their growing inefficiency in many sectors left a deep imprint still far from eradicated from the lands they held longest.

The map of mediaeval Europe can mislead us. The boundaries of states were not only changed frequently, but they commonly lay ill-defined in marshlands of sparse population or great inaccessibility. The states were really the territorial expression of powerful dynasties and families, wheeling and dealing for their own aggrandisement, empowered and ranked in a hierarchy through the Church, whether Roman or Byzantine. Even though language and ethnic characteristics of the people of our own day were already becoming clear, 'national feeling' hardly existed and loyalty to one's lord was the key to community, whereas the territorial success of states was an index of the ability of their king or prince to command allegiance from his lords and the lower orders of society, or even to win the allegiance of another's lords and peasants.

The map of Europe, about 1550, lies at a key point in time: society had recovered from the trauma of the Black Death but lay at the threshold of the invention of printing and the mounting religious struggle in Western and Central Europe. The ability to print in vernacular languages and to spread ideas lay at the root of the emergence of national identity, though the need to stand up and be counted on the religious question was also an important national catalyst. In the north, the Scandinavian kingdoms were marked by the steady growth in the status of Sweden, to reach its peak in the seventeenth century and then quickly fade. Denmark and Norway were to be in an unhappy union from 1380 until 1814, followed by an equally unhappy union between Norway and Sweden until 1905. The kingdom of England had abandoned its major territories in France by the late fourteenth century, though

Calais was to remain until 1558 and the Channel Islands until our own day. The English were deeply involved in the struggle to subjugate their peripheral territories: Wales fell by the early sixteenth century, but the campaign against Scotland culminated in elaborate dynastic manoeuvres and the Act of Union (1707). The long struggle in Ireland was never effectively completed, with the religious issue growing in importance from the sixteenth century.

One of the earliest national identities to crystallise was that of France, led by the policy of unification from the Île de France, while doubtless the long struggle against the Angevin kings of England helped to strengthen the image of France. It was, however, not until the seventeenth century that France began its main expansion eastwards at the expense of the Germans, generating a long conflict of interest in the vanished remains of Lothar's Middle Kingdom (Lotharingia) of the ninth century, making the Rhine a political–geographical watershed. Just as in France the Angevin English had provided a catalyst, so the struggle in Iberia against the Islamic Moors had generated a recognition of community of interest, drawing the petty kingdoms together into the Spanish state, despite early separation of Portugal, perhaps through the environmental contrast of the Atlantic littoral with the bare interior and Mediterranean coast.

Looking at the map of Europe of about 1550, the observer might be confused that the central feature was the large territory of the Holy Roman Empire covering much of Central Europe. German colonisation eastwards had been followed by formation of German principalities, while Slav populations had been either absorbed or displaced by the advancing German peasants until movement was brought to a halt by the Black Death (1348–1351). The boundaries of the Holy Roman Empire had become remarkably fixed by the early fifteenth century. The name of Germany spread across the map represented those lands where people spoke German but it was never a real political–geographical concept, whereas the empire 'of the German Nation' contained many non-German people. The nature of the Holy Roman Empire allowed intense territorial fragmentation, a patchwork of petty states of temporal and ecclesiastical rulers, as well as powerful commercial cities that grouped together in mutual benefit and became the Third Estate among the Electors of the Emperor. This territorial fragmentation was best seen in Central and South-western Germany and in the Rhineland, where particularism and parochialism as well as constant friction between these micro-states remained endemic well into the nineteenth century. Within the empire's borders long and intense struggles for dynastic power – especially the coveted title of Emperor – were waged, though of the many contestants only two were in the end important in moulding the political pattern of Europe. The Hohenzollern family were first to appear significant after the Franconian line took over the Mark Brandenburg in 1415. Through their skill they sowed the germs of the great Prussian state that in the mid-nineteenth century was to become leader of the other German states in the formation of the second *Reich* (the first having been the Holy Roman Empire itself). Already, from the thirteenth century, a little-known family in the south-east marchlands of Germandom – the Habsburgs – were building the foundations of an empire that was to become the classical example of the dynastic expression, a vast polyglot domain stretching far beyond the eastern boundaries of the Holy Roman Empire. The struggle for leadership of the German people and Germandom between these two powerful dynasties was to endure into the opening of the present century.

Italy, into which the Holy Roman Empire still extended until the seventeenth century, was a mass of petty states, including considerable ecclesiastical territories (some under direct papal rule). The great age of the commercial empires of Venice and Genoa was in decline by the sixteenth century and it was the kingdom of Naples that formed a core for the later unification of the peninsula.

In the mid-sixteenth century, the Ottoman Turks were near the peak of their gains in Europe: all the Balkans had fallen to them and they held much of Hungary, while princes of the Rumanian lands served as their vassals and the Black Sea littoral to the Russian Don was in their hands. They were ultimately to reach the gates of Vienna (1683) before the tide turned, saving Christianity, and a long slow decline began as the Habsburgs pressed them back to the Lower Danube. The last long Ottoman rearguard was aided by the jealousies of European powers, for each feared that the collapse of the *Porte* – the Sick Man of Europe – might precipitate a scramble for the spoils in which they themselves might lose out.

It is often overlooked that one of the most powerful European states in the sixteenth and seventeenth centuries was the Polish–Lithuanian Commonwealth, a personal union of the two princedoms after 1386. The Polish–Lithuanian princes saw themselves as the banner-bearers of Roman Christianity, seeking to convert not only the heathens of the 'Wilderness' but also Orthodox Christians. It was indeed the king of Poland who saved Europe from the Turks by their crushing defeat inflicted at Vienna in 1683. For a long time, Poland greatly overshadowed Russia, but ultimately the work of the princes of Moscow in uniting under them other Russian princes (especially in the sixteenth and seventeenth centuries) produced a powerful state whose strength lay more in its spatial character than in its organisational ability. As Russia, Prussia and Austria grew in strength, Poland – riddled by internal dissent and social malaise – softened and weakened, finally to be dismembered by these powerful neighbours in a few brief years in the late eighteenth century.

The unity of the Church of Rome and its role in dispensing grace and favour among the rulers of Europe were rudely shaken by religious turmoil in the sixteenth and seventeenth century. The Reformation and the counter-Reformation added to the dogmata of Rome and Byzantium a third set – Protestantism in all its shades. The tenet *cuuis regio, eius religio* (suggesting that temporal princes decided the religion of their subjects) left, when all the trouble and dispute died down, a pattern of denominations to influence subsequent political–geographical relationships of the nation-states that blossomed in the eighteenth and nineteenth centuries.

The Age of Absolutism

The eighteenth century was the age of the absolute monarch, when the power of temporal princes reached its zenith and they might say with a bitter truth *l'état c'est moi*. New cultural movements were at the same time, however, studying language, folklore and anthropology in a light that raised the dignity of the long-ignored masses to a new level. The map of the Age of Absolutism (around 1740) showed a finally and reputedly United Kingdom in Britain. Sweden was still in its heyday, though already in decline in the southern Baltic, and Norway and Denmark were linked in their unhappy union. In the south-east, the Ottoman Turks were in retreat before the Habsburgs, whose desire to repopulate captured but devastated lands was bringing a new complexity to the ethnic quilt of the Danube basin. In the east an emergent and rejuvenated Russia was pressing back the Turks as it spread southwards and eastwards, while it was also taking lands from Sweden in the north. Poland, a mere weak vestige of its former glory, was about to be carved up between the new powers: a Prussia whose 'armies were its frontiers', the confident successors of Petrine Russia, and an ebullient and still successful Habsburg Empire. A powerful France, despite its social tensions, had pressed to the Rhine and was about to make the great convulsion, the Revolution, that was to recast Europe in the surprising Napoleonic episode, a last near-successful bid for continental hegemony. Spain and Portugal, soporific after their great imperial adventures, lay in their modern boundaries, but Italy was still a conglomeration of petty states preyed on by dynasties – the Bourbons, the Habsburgs, the House of Savoy and others. The Swiss Cantons had exerted their freedom and had left the Holy Roman Empire in 1648. The Netherlands, too, had established independence in a massive struggle against the Spanish Crown, where they had made good use of the natural terrain of marsh and river. Germany was still excessively fragmented into a mass of petty states of divided loyalties, embroiled in the long struggle between Habsburg and Hohenzollern for leadership, though the coveted imperial title had become unreal. The millennium of the Holy Empire was fast drawing to a close.

Napoleon and After

Following the French Revolution and the kaleidoscopic changes of the Napoleonic Wars, Europe was reordered in 1815, to be set in a pattern that was to last for virtually a century. The greatest territorial gains had been made by Prussia, though it was not until the war of 1866 that its eastern and western territories were united through the incorporation of Hanover (other gains also expanded Prussian territories in North-west Germany in 1867, 1876 and 1890). Nevertheless, though the Holy Roman Empire had been replaced by a loose German Confederation, *Germany* remained largely a name on a map rather than a political reality, even though the number of political units had been reduced from 360 to 35 states and four free cities. The Habsburg Empire – to be the Dual Austrian–Hungarian Monarchy from 1867 – reached its greatest extent, having made territorial gains in Italy and on the Adriatic littoral, though the Italian lands were to be lost in mid-century. Russia had gained Finland in 1809 and Bessarabia in 1812, whereas its position in Poland, despite theoretical Polish 'autonomy', had been consolidated after 1815. Italy still remained a collection of petty states, mostly engaged in conflict against

France and Austria, but ultimately to reach unity in the 1860s. The Congress of Vienna had created a United Netherlands, a non-starter union of incompatible Catholics and Protestants, Walloons, Flemings and Dutch, which quickly (by 1831) had disintegrated into the Netherlands, Belgium and Luxemburg.

The two major features of the political geography of nineteenth-century Europe were, first, the continued decline of the Ottoman Empire, with new states emerging: Bosnia-Hercegovina taken by Austria (1878), independence for Serbia and Rumania in 1878, along with Montenegro, while Bulgaria also made the first moves to independence, and Greece had begun its march to the same goal in the 1820s. Secondly, the new vigour imparted to the German lands, with the achievement of economic unity in the *Zollverein*, though progress on the political front was slower. The growth of German economic well-being brought a critical change in the demographic balance in Europe: in the 1860s Germany for the first time exceeded France in population, so swinging the balance of manpower to Germany's military advantage. By careful moves, Prussia steadily gained in power and status among the German princes, particularly through the successful war against France in 1870 (in which Prussia gained valuable Lorraine and Alsace). Capitalising on this victory the Prussian king took the leadership of the German states and drew them, except Austria, into a new empire – the second *Reich* – with himself as emperor. At last one *Germany*, in political–geographical terms, appeared across the map of Central Europe and rapidly became one of the world's greatest economic and military powers.

Despite several petty wars, the nineteenth was a century with sufficient peace to allow major demographic and economic change to take root. Population growth, aided by advances in public health, was vigorous and sometimes outpaced the economic revolution that had begun in England and drifted eastwards across the Continent. New systems of transport – railways and canals – and a new factory-based economy arose within the political–geographical framework cast by the Congress of Vienna in 1815. In the ferment of ideas first laid in the eighteenth century, new concepts of community and identity arose. A vigorous interest in vernacular languages, in folk traditions, in the differentiation in the types of man, strengthened by a widening horizon of literacy and the ability to disseminate knowledge through printing, all added to a new view of allegiance and community – nationalism. Loyalty to a dynasty waned as loyalty to one's people, one's nation grew. Nationalism had helped Poles retain their identity and resist integration in the imperia that had torn their country apart; nationalism had nurtured Greek and Slav struggles against the hated Turks; but nationalism had equally encouraged the Magyars to magyarise their Slav subjects and the Prussians to press germanisation on reluctant Poles. These new feelings of national identity, glorified and respected, began to undermine the credulity of the dynasties (nowhere more so than in the Habsburg lands, where one group was played off against another), though some dynasties sought to identify themselves with the new nationalism – with success by the Hohenzollerns in Prussia and the *Reich*, less so by the Romanov Tsars of all the Russias.

The First World War and its Aftermath

Seized by a dogmatic nationalism and an uncritical patriotism, forgetting how armed conflict had been changed by the technological advances of industrialisation and the creation of popular mass armies, the great European powers – in the most inept and arrogant stupidity – threw themselves into open conflict in 1914, abandoning what seemed to be an accelerating prosperity in a world under European hegemony. They themselves ended the European Age, whose supremacy and hegemony were never again to be so clearly defined. A new map was drawn in 1919–1920 – no longer to respect the wishes of dynasties but to accept those of 'self-determination' of mass national identity.

In Russia, a revolution as fundamental as that of 1789 in France threw the country open to prey from adjacent countries: Finland, Estonia, Latvia and Lithuania freed themselves from the Russian yoke; Poland seized large eastern territories, though only a small part of those coveted by Polish greed; while Rumania took away Bessarabia. The Habsburg Empire, where moderate requests from several groups for some form of autonomy had been ignored early in the century, now disintegrated into the Succession States of national groups taking advantage of a vague and not necessarily apposite American-inspired 'self-determination'. Poland reappeared, carved from Prussian, Austrian and Russian territory, but including large numbers of non-Poles and vested with a frightening array of frontier problems, notably in industrial Silesia, in the unworkable corridor to the sea and in the unfortunate 'free city' of German Danzig. Hungary's borders were drawn tight 'until the pips squeaked' to allow for the aggrandisement of a new Rumania, for the creation of a Czecho-Slovak state and a South Slav state, Yugoslavia. Besides losing territory to Poland and to France and some snippets to Denmark, Belgium and Czechoslovakia, Germany was reorganised internally: the Hohenzollern dynasty, like the Romanov and the Habsburg, was swept away. Because no European power could countenance a rival controlling the narrow waters that gave access to the Black Sea, a tiny Turkish territorial remnant around Istanbul was left.

The neutrals – Holland, Norway, Sweden, Switzerland, Spain and Portugal – came out of the 1914–1918 war territorially unchanged. Denmark had a small territorial gain from Germany, just as Belgium also gained (including a customs union with Luxemburg). France returned to the Rhine by taking back Lorraine and Alsace, while the Saar was under French control until 1935, when it returned to Germany. Civil war in Ireland resulted in the formation of the Roman Catholic Free State (later Republic), with the six Protestant counties of Ulster remaining in the United Kingdom, a tension situation that suppurated into violence in the later 1960s. Italy emerged resentful, with a pocketful of tiny gains (including the Trentino from Austria), while 'Rump Austria' had hardly vitality enough to survive and was denied the right to merge with Germany, a move that might have been for both the economic and political salvation in the years to come.

With an exceedingly complex ethnic quilt – a product of the times before

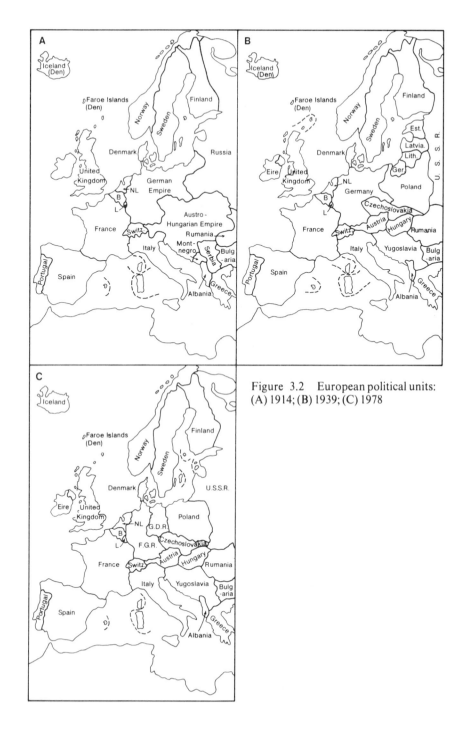

Figure 3.2   European political units:
(A) 1914; (B) 1939; (C) 1978

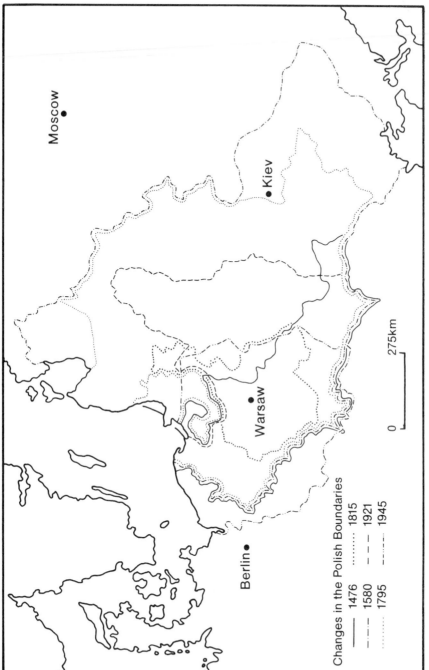

Moscow

Kiev

Berlin

Warsaw

Changes in the Polish Boundaries

——— 1476	·········· 1815
—·—·— 1580	— — — 1921
·········· 1795	—··—··— 1945

0 ⊢———⊣ 275km

Figure 3.3 Changes in Polish boundaries illustrating the vulnerability of a state without natural frontiers

'nationalism', when people moved at the whim of a dynasty – it had been hard after 1918 to define tidy national states without substantial minorities. Some of the states created in the carve-up were without the resource potential ever to reach a reasonable viability, whereas the new boundaries cut across the spatial pattern of the economic life of a prosperous Europe abandoned in 1914. With small nation-states trying to create an identity, with disgruntled minorities, and economic disruption being countered by economic nationalism, nationalism itself became madly uncompromising. Powers that might have held the balance quietly disarmed and put their trust in a flabby and toothless League of Nations that had failed even its initiation. The Americans, whose ideas had so forcefully moulded the new map, retreated introvertly into their own continent, applying the policy of 'isolationism'. The Germans were particularly resentful of the *Versailler Diktat*. Equally, resent smouldered in Danubia and in the eastern marchlands, as well as among the disappointed Italians at real or imagined injustices.

With so many unsettled issues, friction was common throughout the interwar years, increasing after 1933 when German policy sought to revise the Versailles treaties and to recover lost territories. German territorial expansion began successfully – in 1938 came the *Anschluss* with Austria, quickly followed by the absorption of the German 'Sudetenland' districts of Czechoslovakia. Early in 1939 came the Protectorate over Bohemia and Moravia and the reoccupation of the Memelland (lost in 1920 to Lithuania). The Second World War came when, late in 1939, Germany attacked Poland (which like Hungary had already joined the Germans in taking spoils from a prostrate Czechoslovakia). By careful preparation and novel tactics, the German *Wehrmacht* quickly overran large areas of Europe, finally committing Napoleon's fatal error of overextension by attacking the Soviet Union in 1941. Despite facing an immense war-making potential in Britain, the United States and Russia and their lesser allies, and despite their surrounded position with its modest natural resources, Germany came near in *Grossdeutschland* to the hegemony of Europe. After four years, failure to eliminate Britain and to neutralise Russia brought defeat and laid the ground for a new and as yet ill-comprehended Europe.

The Overseas Imperia

No review of the political geography of Europe can stand without reference to the many imperial and colonial ventures that carried European influence far across the world. The emergence of Europe's nation-states was much influenced by this development. Even the states of Classical Greece and the Roman Empire had spread into Africa and Asia, whereas Norse adventurers settled and colonised in Iceland and in Greenland – probably even in North America. Having taken the initiative in oceanic exploration, Spain and Portugal, blessed by the Pope as he apportioned the world between them, sought riches in Africa and in the Americas. The seventeenth century saw active exploration and colonisation by the British, French and Dutch, but by the early eighteenth century Dutch seapower was already in decline, though the Dutch succeeded in retaining their colonies in the West Indies and their valuable

East Indian territories. Much initial colonisation by Britain and France was in North America, where their European rivalry was translated into colonial conflict.

Despite extensive French colonisation in the St Lawrence and lower Mississippi, the British were ultimately to be victorious and in Canada came to incorporate a substantial French population. Yet, Britain's American colonies of New England and Virginia, resenting too tight a hold from the European homeland, were to revolt and establish an independent United States by the end of the eighteenth century. In constant rivalry with France, Britain further spread its holdings in the West Indies and the Atlantic islands and it became supreme in the eighteenth and nineteenth centuries in the Indian subcontinent, besides colonising Australia and New Zealand and gaining wide African territories (including the Dutch Boer Cape Colony). The Spanish and Portugese colonies, initially images of home, in South and Central America grew apart from the Iberian way of life as its fortunes declined, and from the early nineteenth century were in rapid succession to establish a series of weak states of large native and half-caste populations ruled by an expatriate Iberian aristocracy, supported later by financial gain from British, French and American investment.

European penetration of Africa from the late fifteenth century onwards had been from the coasts. One of the important items of trade was slaves for the rich plantations of the New World, though the trade was even more ruthlessly pursued by Arab traders in East Africa for the Asian and Middle Eastern market. Apparently less rich in minerals than the Americas (despite many legends) and more impenetrable ('the Dark Continent'), it was not until slaving raised European disgust and Church missionaries began to search for souls to convert that exploration began, coinciding with a fever of imperial aspirations among the European nations late on the colonial scene. The late nineteenth century brought an undignified scramble for Africa: Portugal clung to its vast territories on the east and west coasts of southern Africa, and Britain's ambition to expand in the east, to achieve an all-red line from the Nile delta to Cape Town, as expressed in Rhodes's Cape-to-Cairo railway concept, was thwarted by German aspirations in what later became Tanganyika (Tanzania). Britain also expanded along the steaming coast of West Africa, whereas France took much of North Africa and spread across the Sahara to reach the West African coast. Few pickings remained for the Germans, though the King of Belgium received the vast Congo basin because the larger powers could not agree among themselves on its apportionment. Almost too late, some Italian colonies were established (though the first attempted conquest of Abyssinia was a disastrous failure), while several small Spanish holdings remained.

European attempts to subjugate China failed, though important concessions were wrung from the declining Manchu dynasty. In Asia, the major European colonial effort, apart from the British in India and the Dutch in Indonesia, was the Russian colonisation of Siberia, brought to a halt when Chinese Manchuria had been penetrated and a clash with an emergent and partially europeanised Japan ended in defeat in the war of 1904–1905 that signalled the end of European success in Asiatic imperial junketings. The search for a north-east and north-west passage to China in the sixteenth century had attracted European interest to the Arctic, where increasing activity, conditioned much by sheer adventure, came to be concentrated from the mid-nineteenth century onwards. The search for an imaginary but immense southern continent had led to the discovery of Australia and New Zealand and was followed by penetration into the truly south polar regions in the nineteenth century. International rivalry, fevered by nationalism, brought desperate attempts to be first to reach the poles.

The Imperial Age (1650–1914) had provided Europeans with an outlet for adventure, for wealth and for colonisation. It was a way to augment home resources and to provide things that Europe could not – such as spices, sugar, cotton or gold and silver. It provided an escape from supposed overpopulation at home – such as the great migrations that arose in the late nineteenth century – or from standards and conventions in the homeland that were no longer acceptable (the motivation of the Pilgrim Fathers). At the same time, especially in the mid- and later nineteenth century, Europeans came to feel they were bringing a better life – religion, education and material progress – to the natives, for whom they developed a patriarchal attitude of the 'White Man's Burden'; though wherever the natives stood in the way of imperial aims they were brutally brushed aside. In an age of nationalism or dynastic rivalry, it was a way to carry the power struggle beyond the homeland. In this way, a potentially devastating conflict, for example, between France and Britain in the Channel, could be defused by privateering in the West Indies. By the end of the nineteenth century, possession of colonies and the imperial urge had become a symbol of national virility and a mark of strength, so that the British, for example, derived immense comfort from their empire 'on which the sun never set'. The European territorial possessions overseas were also a great commercial incentive and much industrial growth, inventiveness and prosperity in Europe derived from these colonial involvements. Without such stimulus, growth at home would have been slower and without the involvements the European institutions, languages, religions and cultures would not have been dispersed around the world out of which the essential elements of a world culture have evolved.

By the twentieth century many European colonists had begun to discover the separate identity so long ago sensed by the American colonists – Canada, Australia and New Zealand emerged as European nations beyond Europe and the new situation was catered for by converting the British Empire into a Commonwealth. Equally, native people discovered or rediscovered their identity and sought independence. The British said they expected 'the ripe fruit of the colonies to drop ultimately from the mother tree' and aimed to create native state organisms in Africa and in Asia; the French sought to counter such independence by encouraging all subjects to become Frenchmen; the Dutch ignored the portents and the Portuguese hardly believed their integrationalism would lead to such demands. The German colonies had been eliminated by the other powers in the First World War. The Italians held mostly desert, and their conquest of Abyssinia in 1936 was the last great European imperial fling. In Asia, the Japanese victories in the Second World War reduced the status of European powers in native eyes, and the critical gap between the Japanese collapse and re-establishment of the old colonial regime in 1945 was the precipitant of independence in Indo-China and in Indonesia as well as Burma; while encouragement of Indian nationalism by anti-British powers helped bring forward the day in India. Encouraged by Asian success, Africa followed in the 1960s, and simultaneously many territories sought their freedom, whether or not they were politically and economically ready. Yet everywhere the mystic attraction of European political and economic concepts and artefacts remains strong: in Japan, never anything but independent, even European music has been accepted; and South-east Asians prefer cowboy films to the products of Indian or Japanese studios. Although many European influences may well endure, some newly independent states find democratic institutions and methods hard to nurture in an atmosphere of economic and social turmoil, when totalitarian systems appear to achieve results more speedily.

Reference

1. A useful companion to this chapter is a good historical atlas. Recommended are Muir's *Historical Atlas – Ancient, Mediaeval and Modern*, Philip, London (1962) or *Westermann's Grosser Atlas zur Weltgeschichte*, Westermann, Brunswick (1972).

4 Europe's Population

Europe (excluding the Soviet Union) contains about 15 per cent of world population, though this proportion has already fallen from a slightly larger share reached during the remarkable explosion of population in Europe during the nineteenth century, and by the year 2000 it is expected it will have further declined to well under 10 per cent. This is primarily the result of the fall in the rate of natural increase in Europe, whereas that of many Third World countries remains high. Europe has made a major contribution to world population in the extensive migration of its people to other continents – if we count together European population and populations elsewhere of European descent, the share of total world population rises to nearer one-quarter. People of European descent cover much of the Soviet Union (where 80 per cent of the total is of Slav or other European descent), North America, South and Central America, Australasia and parts of southern Africa. These groups outside Europe have in general already developed such distinctive societies and cultures of their own that they are no longer to be considered simply as expatriate Europeans.

Table 4.1 shows that the regional share of population varies widely: clearly very important are Western and Central Europe, whereas in relation to their area, Iberia and Scandinavia have relatively meagre shares. The share of South-east Europe is likely to rise appreciably, since population growth is still vigorous there. If we consider the distribution in terms of political units, then West Germany, the United Kingdom and France clearly dominate, or we may express it in the contrast between the European Community of the Nine (257.2 million) and the European members of Comecon excluding the Soviet Union (105.7 million), 54.7 per cent and 22.5 per cent of the total, respectively. In manpower terms, the range between the population of the largest political unit (West Germany, 62.0 million) and the smallest (for example, Monaco, 24 000; Holy See (Vatican), 1000) is extremely great. Such purely statistical comparisons, although interesting, give us, however, little picture of the true distribution of population.

From the confusing pattern of tints and dots on a population distribution map certain striking features can be discerned. The most notable are the two main belts of high density settlement (generally more than 100–150 persons/km²), one running east–west along the northern edge of the Central European uplands, and the other north–south along the general alignment of the Rhine valley and, though interrupted by the Alps, into the Italian Peninsula. Beyond these two belts, clusters of population occur in various favoured areas and around some of the larger cities. Even at a glance, the density of population is

Table 4.1   A synopsis of Europe's population

Total European population (excluding U.S.S.R.)	470 000 000
Western Europe (population as % of Europe, 16.21):	
France	52 507 000
Netherlands	13 541 000
Belgium	9 804 000
Luxembourg	342 000
	76 194 000
West Central Europe (16.18%):	
Federal Republic of Germany	62 041 000
Austria	7 528 000
Switzerland	6 481 000
	76 050 000
East Central Europe (16.04%):	
Poland	33 691 000
Democratic Republic of Germany	17 166 000
Czechoslovakia	14 686 000
Hungary	10 458 000
	75 379 000
Mediterranean Peninsulas and Islands (13.89%):	
Italy	55 361 000
Greece	8 962 000
Cyprus	641 000
Malta	323 000
	65 287 000
United Kingdom and Republic of Ireland (12.47%):	
England and Wales	48 750 000
Scotland	5 229 000
Northern Ireland	1 528 000
United Kingdom	55 507 000
Republic of Ireland	3 086 000
	58 593 000
South-East Central Europe (11.34%):	
Yugoslavia	21 153 000
Rumania	21 029 000
Bulgaria	8 679 000
Albania	2 416 000
	53 277 000
Iberia (9.36%):	
Spain	35 225 000
Portugal	8 735 000
Gibraltar	27 000
	43 987 000
Scandinavia (4.71%):	
Sweden	8 161 000
Denmark	5 045 000
Finland	4 682 000
Norway	3 987 000
Iceland	215 000
Faroe Islands	40 000
	22 130 000

Source: United Nations Yearbook.

clearly related to relief, for the higher uplands and the mountains stand out generally as representing the more sparsely settled areas, usually with densities as low as 10–25 persons/km².

The east–west belt may be considered as starting in North-west England and crossing the English Midlands to South-east England, where densities are well in excess of 150–200 persons/km² in many parts and include the large conurbations around Manchester and Liverpool, Birmingham and the London area. Across the Channel, this belt encompasses much of the Netherlands (including the unusual annular conurbation of the *Randstad Holland*) and Belgium, where densities are generally higher than in England. Eastwards it passes into the landscape of town and suburbanised country of the Rhenish–Westphalian industrial district with densities of over 200 persons/ km²: here the town is melting into the countryside with attendant planning problems. Still further eastwards, high rural densities mark the excellent loessic soils of the *Börde* country, amid which are dotted many small industrial towns, with significant clusters of people in the mixed agricultural–industrial economies around Bielefeld, Hanover and Brunswick. The belt is now cut by the border between the two German states, though densities on the east have increased less than on the west. Nevertheless, the basin of the Elbe and Saale rivers, where rich loessic soils are underlain by brown coal, have densities well over 100 persons/km², locally much higher around industrial towns like Halle and Leipzig. Densities over 200 persons/km² occur in industrial Saxony centred on Karl-Marx-Stadt and also around Dresden. Political events since 1945 have changed the character of this belt on the good loessic soils of Silesia, where the expulsion of the Germans and resettlement by the Poles has left the countryside and most smaller towns with populations well below the 1939 level. Densities have, however, been maintained in industrial Upper Silesia, where most towns are well above their 1939 population. The belt continues across southern Poland, where prewar rural overpopulation has not been fully eliminated, though densities are considerably lower in the poorer agricultural country of the San basin. East of Przemysl the belt broadens into the Ukraine (though densities are again lower, especially on the drier southern edge) and ends eventually in the coalfield of the Soviet Donets basin.

North Germany and Poland beyond the main east–west population belt have only mediocre farming conditions, marked by densities of around 50 persons/km², and there are considerable forests. Distinct clusters of people are associated with the main towns – in Germany, Hamburg and the lower Elbe, Bremen and Kiel as well as the big cluster of Berlin, whereas in Poland Warsaw and Lódź are notable. In central Poland, rural densities on only mediocre soils with peasant farms nevertheless generate about 100 persons/ km², but this countryside includes the now partly industrialised district extending south from Warsaw towards Radom and Kielce. The rural districts east of the Oder river absorbed by Poland in 1945 are still modestly settled despite official resettlement from the interior to replace the expelled Germans. Fewer than 25 persons/km² over considerable tracts of country occur here, well below the 1939 level, though industrial towns mostly exceed their prewar populations.

The north–south belt diverges from the east–west belt on the lower Rhine

near Duisburg and extends south along the river, with some very high densities occurring locally, particularly in the gorges where the vine supports substantial agricultural populations. There are several large and important
towns, but surprisingly high densities are found in the uplands east of the river
owing to the presence of many small industrial towns, besides recent residential development. The Rhine–Main area is marked by labour-intensive industry and vine cultivation able to support populations of well over 200
persons/km². The remainder of the agriculturally prosperous rift valley,
away from the immediate flood levels of the river, also has a dense population, generally thicker on the eastern side, where there are several large industrial towns. The well-settled rift floor contrasts with the modestly peopled
forested uplands on either flank. Associated with the Rhine population axis
are the densely settled industrial countryside of the Neckar basin, which in-
cludes the cluster of industrial communities around Stuttgart, and the scarp
foot industrial towns, as well as the meandering Main valley. To the west,
routes lead to the Saar coalfield, with its mining villages and industrial towns
that give densities exceeding 200 persons/km² along the Saar river itself.

South Germany, away from the main north–south Rhenish population
axis, has country with widely varying population densities. In the better agricultural land of the rich vales, often with vine cultivation, amid scarplands a
density of more than 100 persons/km² is not uncommon, whereas on the dry,
exposed dip slopes with poor backward farming and extensive forests seldom
even half this density occurs. Well over 50 persons/km² occur on typical
farmlands (that include special cultures such as hops) of the gravelly terraces
and Tertiary hill country of the Alpine Foreland. Significant clusters occur
around the main towns where industry has grown rapidly in the past thirty
years, but bare thinly settled patches are found on moor and heathlands.

The Swiss *Mittelland*, from Lake Geneva to the Lake of Constance, comprising a part of the north–south belt, has a pattern of well-settled countryside
with small industrial towns, but the belt is interrupted by the massive barrier
of the Alps, to the south of which a densely settled countryside lies in the
Plains of Lombardy, where modern labour-intensive industry is superim-
posed on intensive commercial agriculture. Here, among many big and successful towns, lie the 'million' cities of Milan and Turin. North of the Po,
densities on the fertile and well-watered intensively farmed country rise to
well above 200 persons/km², whereas densities on slightly poorer land south
of the river mostly slightly exceed 150 persons/km². This belt may also be
regarded as extending along the northern flanks of the Apennine peninsula,
with a particularly intensively cultivated and well-settled coastal strip between Ravenna and Pescara, while the poor and rurally overpopulated
country of the 'heel' (around Bari and Taranto) can also be included in this
belt. Around Foggia and Gargano in dry limestone country, however, density
falls to a mere 25–50 persons/km².

In the Italian peninsula, where the rugged and mostly inhospitable Apennine backbone repels people, though favoured mountain basins are well settled, the coast attracts settlement. Although the east coast may be regarded as
a continuation of the main north–south population axis, the west coast is
marked by several significant clusters arising from a combination of good

agricultural land and industrial development. Concentrations of people occur around Genoa, Pisa and Florence (where considerable areas have more than 200 persons/km²), Rome and its environs. Particularly striking is the intensive farming on the rich volcanic soils around Naples and Vesuvius, with more than 500 persons/km², whereas some small but dense clusters of settlement occur around the coasts of Sicily on favoured coastal lowlands.

Accessibility and economic and social conditions reflect the pattern of population distribution outside the main belts. Scandinavia may be considered as Europe's 'empty quarter': large areas of interior Norway, Sweden and Finland, particularly in the north, are extremely sparsely settled, apart from mining and industrial centres, and some of the mountains are virtually uninhabited lands. Regional policy in Norway seeks to develop northern growth points with a wide economic basis. In coastal Norway, population occurs in small clusters on suitable sites for limited farming by the fiords and for fishing on the islands. Overall densities seldom exceed 25 persons/km², except in the south in the more extensive and fertile country around Oslo, where densities above 50 persons/km² are found. In Sweden, the more attractive agricultural central lowland lakes and the south, with its many manufacturing and market towns, have densities of around 10–25 persons/km², but the most southern district of Skåne, with the best farming conditions and vigorous industrial growth, has densities around 50 persons/km². Other concentrations are found near the main towns, like Stockholm, Gothenburg, and Uppsala. The sparse population of Finland, boosted by a large refugee influx from territory taken by Russia, lives mostly around the coast, particularly in the south on late and post-glacial marine and lacustrine soils, and in the better lake country. Notable clusters lie in the environs of Turku and Helsinki. Denmark is generally more densely settled farming country, with well over 50 persons/km² on the better eastern farmlands and the islands, where conditions are not unlike elsewhere in the North European Plain. About a half of the national population lives on Sjæland, notably in Copenhagen and its environs.

France is remarkable among the more important European countries for its low overall density, and population distribution in detail reflects well its nature as a land of large and agriculturally rich river basins draining from a less hospitable upland core, the Central Massif. The most densely settled districts are the industrial lands of the north, north-east and the Paris basin, though some closely settled country, including new industrial communities, occurs around Marseilles and the Mediterranean coast, as well as in the Rhône valley. Of the main agricultural lands, the best and most densely settled farming country (densities of well over 50 persons/km²) lies on the moist Atlantic coast, notably in Brittany, whereas in the Paris basin rural density on good farming land is generally seldom above 25 persons/km², although somewhat higher densities occur in the lower reaches of most of the main rivers. In the Central Massif, with its forests, fields and pastoralism, and in the drier eastern farming lands, densities are usually below 25 persons/km², except in the most favoured areas.

The less hospitable interior of the Iberian Meseta is also reflected in the higher population densities of the coastlands, particularly along the well-watered northern coast and on the western Atlantic coast, where the low-lands, often important vineyard areas, leading inland are well settled, with more than 100 persons/km² in places, though the southern part is the least densely populated. The environs of Lisbon and the Porto district, both with growing industry, have densities exceeding 200 persons/km². More than 50 persons/km² occur along the Mediterranean coastlands wherever water is ample enough to support rewarding farming – notable are the industrial and farming lands around Barcelona, between Valencia and Murcia and much of Andalucia. The coastal areas and islands have attracted tourist develop-ments, with consequent considerable seasonal swings in population.

Well-farmed if not very productive countryside, because of soils and alti-tude, covers most of Bohemia–Moravia, where densities vary between 50 and 100 persons/km², though many farmers find at least part-time employment in in-dustry, and there are also sparsely settled areas of extensive forest. Around the main industrial towns are patches of higher density, and the good soils of the upper Elbe (the Polabí) support over 100 persons/km². Dense settle-ment is generated by the many modest industrial towns and villages of upland northern Bohemia and also by the thickly settled coalfield and heavy indus-trial landscape of Czech Silesia. A striking spread of countryside with a rela-tively high density of population for a rural area covers the large mid-Danube basin and extends into Lower Austria, though the principal population agglo-merations are associated with large towns of present or past political import-ance – Vienna, Bratislava and Budapest. Well settled are the Hungarian industrial districts, especially the areas around Debrecen and Miskolc. More thinly settled farming country occurs south of Lake Balaton (where densities fall to about 50 persons/km²) and on the sandy interfluve between the Danube and Tisza. The Tisza–Danube confluence, with excellent farming conditions if backward techniques, in the Banat and Bačka country is well populated, with more than 100 people/km² extending from around the food-processing town of Szeged to south of Belgrade. It continues further south at slightly lower densities in poorer farming country towards Niš and Skopje, associated with the important Morava–Vardar Corridor.

The well-settled plains of the middle Danube are separated by modestly settled mountains from the agricultural country of the lower reaches that extend through Wallachia and into Moldavia on the north bank and in-cludes the broad fertile platform on the Bulgarian bank. Almost everywhere densities exceed 50 persons/km², except in the drier country like the Bără-gan steppe and parts of the Dobruja. Even this modest density was high enough in relation to farming standards to cause acute rural overpopulation in prewar times. Between Bucharest and Ploieşti the countryside has con-siderable industry and densities rise sharply, whereas high density farming population occurs along the Carpathian footslope, another area of con-siderable pre-1939 population pressure.

The Carpathian mountains, with large forests and high pastures, stand out as thinly settled countryside, cut by densely settled valleys extending far into the mountains. In Rumania, the Bihor Massif, with its modest popula-tion density despite some mining communities, contrasts with the well-settled farming country around it in Transylvania and along the edge of the

Pannonian Plain. Nevertheless with reasonable farming conditions and mineral wealth, densities in this massif remain relatively high to considerable elevations compared with the main Carpathian ranges. The Dinaric ranges separate thickly settled farming country of the Sava and Drava basins and the Voivodina from the clusters of population on favoured sites along the Adriatic coast. As these ranges form dry and inhospitable country, population is clustered in the more promising depressions (the *polja*) and valleys, so that with a general density of over 10 persons/km² the Dinaric mountains appear more thickly settled than might be expected, especially in the higher south. Historically this originates from the retreat of Christian population to the safety of the mountains to escape mediaeval Turkish depredations. A similar character is found in the Balkan mountains and in the Rila and Pirin massifs, where large secluded mountain basins proved most attractive retreats, though the broad and fertile Maritsa depression of Thrace has a relatively dense population of virtual market gardeners. The mountains of Greece have a similar pattern of population distribution, though the long dry summer tends to make access to water important in producing clusters of settlement.

The Alps reflect many features of population distribution commonly found in Europe's mountains. There is the decline in the density of settlement with elevation and the importance of sunny slopes in attracting people, who also cluster along the lower and gentler slopes of the valleys (with densities of more than 50 persons/km² and, in some favoured valleys, of well over 100 persons/km²). Everywhere, known avalanche paths are avoided as are areas where there is pronounced cold air drainage in winter. The highest, steepest and most shaded surfaces are virtually uninhabited, though high meadows have a summer occupancy. The drier southern Alps display a tendency for people to gather on the moister middle elevations, often in positions where defence was easy. Historical influences, even different regional cropping patterns and farm economies as well as settlement types, influence the detailed pattern from one part of the mountains to another. In the Pyrenees, the steeper and moister northern flank has a denser settlement than the gentler but drier southern flank, while population is thicker in the moister west than the drier east.

Age and Sex Structure

The population geography of Europe is usually considered in quantitative terms – the spatial distribution of its people – as varying regional densities or settlement patterns, but we cannot ignore the qualitative aspect – the age and sex structure of population and the related questions of birth and death rates and even life expectancy.[1] All these factors have considerable bearing upon regional social and economic development.

In a world setting, the demography of Europe is characterised by the ageing population and by slow growth, the result of the close approximation between birth and death rates, though it is one of the few areas of the world where a reasonably long expectancy of life can be enjoyed. These factors underlie the shrinking percentage of Europe's population in the present explosive world scene. Nevertheless, the demographic experience of Europe

since the late eighteenth century seems to be proving relevant in other continents, especially in rapidly industrialising countries: using this experience,
many demographers believe that the present rapid growth of world population is an unusual phase that will pass, as it already has done in the overwhelming part of Europe.

The model of demographic transition in figure 4.1, based on this European
experience, has been most clearly illustrated by the population history of the
United Kingdom, though it is evident in other countries as well. Generally,
however, the onset of the rapid growth phase has come later towards the east
and south, as it has been linked closely to the incipience of industrialisation.
The pattern varies somewhat between countries, particularly in the degree of
divergence between the birth and death rate in the second and third phases, so
affecting the rate of growth (for example, in the contrast in development between France and England and Wales). The model consists of four phases,
but some demographers consider that a fifth phase may be expected. The
opening phase, which had ended in the United Kingdom by the early nineteenth century but which ended only much later in Germany, for example, has
a high death rate (particularly arising from infant mortality) offset by a high
birth rate, so that population increase is slow. This is a pattern that remains in
rural areas long after change has come to urban demographic patterns. A
modest improvement in public health, even a somewhat better diet, in the
opening period of industralisation brings a fall in the death rate (again mostly
among infants but with some improvement in life expectancy), but the birth
rate remains high. The third phase is marked by continued fall in the death
rate, but also there is a decline in the birth rate seemingly associated with
improved infant survival. As more people enter industrial employment, children become less significant in the family economy and subsequent family size
tends to fall. The fourth phase is noted as the birth and death rates level off,
both at a fairly low rate, introducing again a low rate of population growth in
contrast to the explosive nature of phases two and three. Greater affluence
and higher life expectancy are reflected in smaller families and later marriage.
Since the 1960s the spread of the contraceptive pill and a further delay in
marriage and child-bearing in several parts of Europe raise a query of
whether we may soon expect to see the death rate exceeding the birth rate, artificially depressed by the pill, as large classes from earlier times emphasise
the ageing population. In this possible fifth phase, natural decline in population will begin, perhaps forcing countries to consider how policies might establish a 'steady state' population. As this pattern of transition has moved
eastwards, the trend has been for the gradient of change in phases two and
three to steepen and the duration of the phases themselves to shorten.

Natural increase – the product of the excess of births over deaths and dis-
counting migration influence – varies regionally. In general, country districts
usually show a higher level than towns, but regional variations are underlain
by a complex mosaic of social and economic considerations. In interwar
Europe, natural increase was low in Scandinavia, but also in the United
Kingdom, in Switzerland, Austria and Czechoslovakia, while in France a
slight natural decrease was recorded (considerably greater among native
French than among immigrants). The troubled 1930s were reflected in the

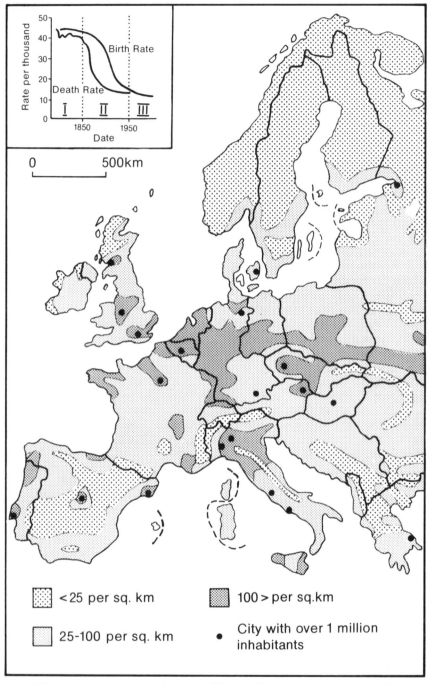

Figure 4.1 Population densities of Europe. Inset: demographic transition

low natural increase in Spain, where a high death rate offset a traditionally high birth rate. Stimulated by social pressures, a high birth rate in the Netherlands was set against a modest death rate to give a natural increase unusually high for Western Europe. In Nazi Germany, government policy to stimulate population growth boosted the birth rate above a death rate that was among the lowest in Western Europe, to result in a formidable if brief natural increase well above the level of the United Kingdom. Eastern and Southeastern Europe, however, showed rates of natural increase well above Western and Central Europe: here a very high birth rate was sufficiently greater than the high death rate to produce a substantial natural increase by European standards.

After the Second World War a crop of governments introduced policies to stimulate population growth, so in Western Europe natural increase was appreciably higher – particularly in the 1960s – than before 1939.[2] The introduction of the contraceptive pill in the late 1960s has been followed by a precipitous decline in the birth rate in countries where it has been accepted. Particularly notable in the early postwar period was the rise in the rate of natural increase in France, where much encouragement through generous welfare payments was given, and in Iberia. The early postwar Welfare State in Britain also helped to boost natural increase. In contrast, very modest levels of natural increase, conditioned by prevailing social and economic conditions, occurred in Austria, Hungary and the German Democratic Republic (where the strong migration of younger people and families to West Germany deprived the state of many potential births). Natural increase in Eastern and South-eastern Europe has remained above the levels of Western and Central Europe, where natural fecundity is less. Whereas high rates of natural increase characterised Poland and Yugoslavia, the rate in Albania has been exceptional for Europe, with a high birth rate set against a modest death rate to show a natural increase more in keeping with levels in Third World countries and twice as high as its nearest European rival.

From the latter 1960s birth rates have tended to drop, sometimes quite drastically, whereas death rates have been edging upwards, not so much through poorer medical care and contrary to a falling infant mortality, but because of a general ageing in most populations. Urbanisation in Eastern Europe, where there is also a generally rising level of affluence, has begun to make children less valuable members of the family micro-economy, setting birth rates on the downwards slope. Most governments have also been less anxious to boost population than in the first two decades after the Second World War. In West Germany, for example, the birth rate of native Germans has fallen below their death rate, in contrast to the still high birth rate among the almost 2.5 million foreign guest workers and their families.

A product of demographic change, the age and sex structure of any population, may be represented graphically as a pyramid, which gives significant clues to the past and future 'manpower' resources. Age–sex pyramids for selected European countries are shown in figures 4.2 and 4.3. The example of the United Kingdom is shown for 1891 and for 1975: it will be seen how it has lost its broad base, characteristic of a youthful and growing population, for the steeper gradient and narrow base of an ageing population. In the case of

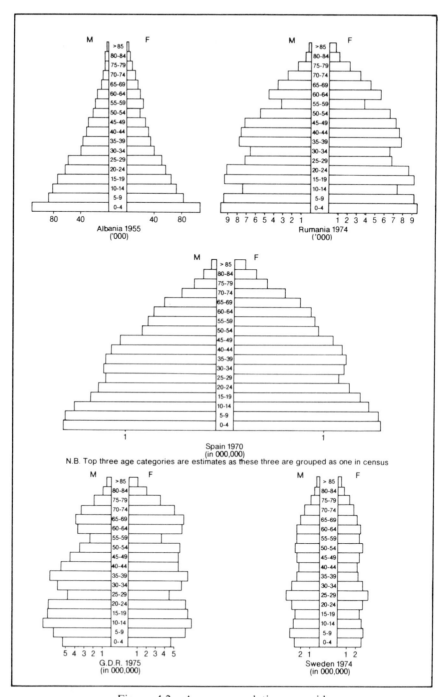

Figure 4.2   Age–sex population pyramids

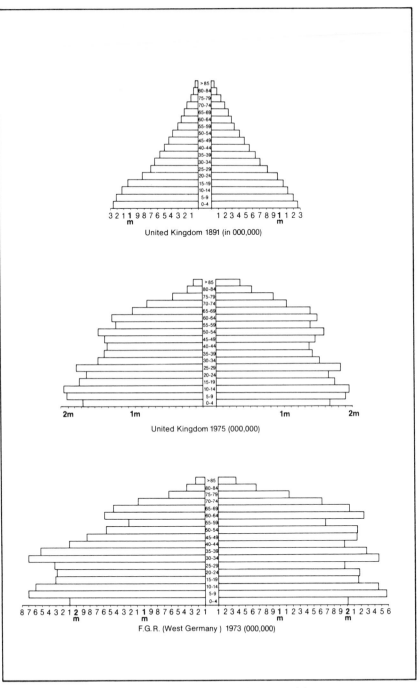

Figure 4.3 Age–sex population pyramids

the Eastern European examples, the element of an expanding and relatively youthful population is retained, at least in part, but a notably youthful population structure is represented by Albania. The population pyramids for the two German states are good examples of the effect of war and other catastrophic events on population: the big bites taken in the sides of the pyramid by the two world wars stand out particularly clearly on the male side, but also apparent is the loss of births in the difficult years of the economic crises of the interwar years. The tendency of the East German pyramid to topheaviness has been accentuated by the vigorous migration of young people to West Germany, whose pyramid has been correspondingly swollen in the lower age groups. The East German pyramid shows the narrow base indicative of a declining population. Most countries of North and Western Europe show the sugarloaf-shaped pyramids of ageing population and the remarkable evenness in the cohorts up to the sixth decade is a biological novelty in comparison with the general world pattern. In Eastern Europe, the big cohorts born in the early postwar period have now passed up the pyramid and the narrower base of most pyramids reflects movement towards the fourth phase of the demographic transition diagram. The long-term implications of these diagrams are smaller and generally declining classes of young people, unless some new stimulus to growth emerges (certainly government policies were not outstandingly successful in achieving this over long periods): this in itself will further slow growth. The shape of the pyramids suggests infrastructural problems in the existing provision of social services (e.g. overcapacity in schools and overmanning in teaching staff, underprovision of geriatric facilities). There will be a smaller annual entry to the labour market, with many economic implications, though there will be a rising reservoir of know-how in the larger older cohorts, which are perhaps naturally more conservative and less innovative (important considerations for future high technology). As a larger proportion of older people form the population, any lowering of retirement age will only increase the growing seriousness of a bigger non-active population and the consequent inflated social welfare costs. Clearly these factors in themselves will greatly change the face of Europe.

Historical Genesis of the Population Pattern

The distribution pattern of population is the product of a long process of differential regional growth or decline, the result of natural demographic processes and migration which may be related to carrying capacity through the different endowment of regions with natural resources and consequent attraction to man. Whereas different rates of natural growth or decline *in situ* have been important, the effect of migration has been of particular significance and has been generated by many different stimuli. Quantitative examination of change is, however, difficult before the regular making of censuses in the latter nineteenth century, though evidence prior to this is usually sufficient to give some general picture of migration trends and other changes.

Migrations and fluctuations in natural change have obviously been operative since the first men-like creatures set foot in Europe, though evidence is too scanty and uneven to give more than a thumbnail sketch for the whole of Europe. Great changes clearly took place in the massive migration of peoples from the fourth to tenth centuries A.D., generated by the collapsing and contracting Roman Empire – certainly much of Europe's ethnic pattern was laid out at this stage, particularly through the restlessness of the German and Slav tribes, but also through incursions by wild peoples from the interior of Asia and the maritime rovings of the Vikings. Most significant for the modern map was the period of active colonisation from the eleventh to late fourteenth centuries under rising population pressure, particularly in Central Europe but also in other parts. There followed a long period of relative stagnation after the depredations of the Black Death, until a new phase of natural population growth and vigorous migration came with the economic revolution that rose in tempo during the eighteenth century and into our own time. In some parts of Europe it is still vigorously in progress, but in others it appears to have lost its demographic if not economic virility.

Charters, place names and an array of historical documents illustrate the extent of the spread of settlement, particularly into the massive forests, notably in Central Europe, in France, in Britain and in Scandinavia. In the latter, it continued longer, though with a sparser population it made less impact in the landscape compared to the great clearings elsewhere. Despite the storminess of the mediaeval period, reclamation and colonisation of coastal and riverine marsh were also made in Western and North-western Europe – it was in this period that the modern landscape of the Low Countries began to take shape. In Britain and in France, this inner colonisation filled out empty space, but in the North European Plain and into the fringes of Danubia, German peasant colonisation spread against a sparse indigenous Slav population, though both Slavs and Germans busily colonised Bohemia. A similar Spanish colonisation also spread against the Moorish hold in Spain. Although much land soon showed itself unsuited to settlement under the then prevailing conditions, nevertheless the pattern of settlement and village distribution had been established in virtually its present form by the mid-thirteenth century.[3]

Considerable movement took place in Danubia and South-east Europe, though its extent and nature is more difficult to assess. One of the most disruptive elements that left an impression lasting into modern times was the spread of the Ottoman Turks, whose ruthlessness and disinterest in economic development combined to force changes in settlement and population distribution. Many Christian Slavs retreated to the mountains, inhospitable and offering little chance of a fair living, rather than become 'sheep' of the Ottoman 'shepherds': the present remarkable densities of population in many secluded upland valleys and basins in the Dinaric and Balkan mountains still reflect this movement. In the Hungarian Plains, the search for safety from Turkish raids produced numerous immense town-like villages as people clustered together for communal action – the Puszta still bears a clear imprint of this period. The Habsburgs in the eighteenth century resettled many areas devastated and depopulated by the Turks with colonists from several quarters of their polyglot empire, so creating an intricate and complex ethnic pattern in the Banat, Bačka and Voivodina.

Collapse of good order in society or bad economic practice also influenced population distribution, often long after the offending events had passed. Overgrazing of mountain pasture, ruthless felling of timber, the free ranging of goats in forest and scrub, all generated soil erosion which drove people away or accentuated coastal marshiness and consequent endemic malaria that deterred settlement. Such influences have greatly affected the pattern of settlement and the distribution of people in the Italian Peninsula, in the Spanish Meseta and in the Balkan and Greek lands. Yet today we can find little trace of the impact of the Thirty Years' War in Germany, though at the time population in war-torn districts was often reduced by more than one-half. People show a remarkable resilience to return and settle areas unless they have suffered lasting environmental damage.

Great changes in population distribution were recorded during the nineteenth century. Although we usually associate these with industrial development in Western Europe, particularly far-reaching changes were associated with the decline of the Ottoman Empire. For example, as settlement became safer in Wallachia and Moldavia peasants spread into lands once dominated by Turkish raiding parties, while people drifted back to the plains from the Balkan mountains, filling lands which Turkish *čifliki* had been uninterested in cultivating, and the settlement of devastated lands by the Habsburgs (and others as well) has already been noted. With a renewed interest in farming, Central and Eastern European princes and nobility encouraged land clearing and reclamation by peasants, though the main effort was made in Prussia. Improved technology also made reclamation of riverine and coastal marsh easier in the second half of the century, with much activity in the Netherlands bringing consequent spread of settlement.

The main change in patterns of population distribution came, however, with industrialisation, first in Britain and then spreading into Western Europe. Factory industry attracted workers from an ever-widening range of countryside and contributed to massive urbanisation. The impact of these changes felt in Britain in the 1840–1860 period was later in Germany, for example, with really fevered development generated after 1870, whereas not until this century was it felt in East Central and Eastern Europe – the revolution is at present in full flood in Rumania and Bulgaria, as well as in Yugoslavia. Changes were, however, dramatic – for example, the emergence of great industrial towns and industrialised areas in once poorly settled country, as in Lancashire and Yorkshire in the 1790–1850 period or the growth of crowded Clydeside slightly later along once quiet river banks. After 1860 similar experience was had in Germany, with explosion of old towns and the mushroom growth of new ones, most clearly demonstrated on the Ruhr coalfield. The effect in Britain was a marked shift in the centre of gravity of population – in England, the West Country and parts of Eastern England, with their rich old guild towns and well-settled farming country, were overtaken as a population focus by the new non-guild towns with their industry and mining in the north, while coal mining in the South Wales valleys transformed quiet backwaters to pulsating arteries of trade and people. Although less intense, similar change was seen in northern France and the Low Countries, in the Saar and in the Elbe–Saale basin (largely after 1900), whereas on a less striking scale Bohemia, Czech Silesia and the Upper Silesian areas witnessed the same change also around the close of last century.

The most notable feature in population distribution from the early nineteenth century onwards has been the increasing agglomeration expressed in the growth of towns and the emergence of conurbations with a widening gap between density of settlement in the urbanised areas and in the poor and mediocre farming lands, where the frontier of settlement has in some instances retreated. Europe may be considered the 'urban continent'. Agglomeration has been largely achieved through migration into the economic growth areas, whose natural growth (excess of births over deaths) has often been substantially below poorer rural areas. Migration away from country districts has, if maintained strongly enough and long enough, resulted ultimately in local population decline, as seen in Ireland and in Highland Scotland, but also in many upland and mountain areas of Europe. Such migration leaves the older folk behind and an ageing population has in some instances left communities completely unviable. At the other end of the scale, intense agglomeration – either as vast urbanised tracts or as massive towns – can produce infrastructural difficulties that result in excessively expensive cost structures. A late twentieth-century planning dilemma is how to control these developments at both ends of the population spectrum.

Change in population distribution has also arisen in some instances from political–geographical change. Even by the sixteenth century, there was politically inspired settlement: the 'plantations' of Scottish and English settlers influenced population distribution in Ireland; on the European mainland, religious strife also produced substantial migration (for example, the Huguenots), whereas Habsburg political–geographical considerations marked the establishment of the settlement of the military frontier against the Turks. In the age of nationalism, boundary and territorial changes were almost invariably accompanied by movement. Considerable migration arising from boundary changes was generated by the Franco-Prussian War, but the scale after the First World War was much larger, and an effective and well-executed exchange of population between Greece and Turkey in the early 1920s solved a long-standing ethnic problem (1.4 million Greeks returned to Greece, whereas 350 000 Turks left Greece for Turkey). The change was much more Draconic after the Second World War, since when the ethnic and population distribution pattern of East Central Europe has been drastically recast through Soviet action. A millennium of German settlement was destroyed by flight and expulsion, leaving large tracts of countryside almost uninhabited for many years, but considerable exchanges and movements took place between other countries as well, involving in all about 20 million people.

As strong and as complex as the internal migrations in Europe may be – whether intra- or international – a most impressive feature has been movement from Europe to other continents. These are almost unique and their nearest rival – the carriage of Africans to the Americas – was also generated by Europeans, who also encouraged the movements of labour between India and Africa and from China to South-east Asia, which have had a powerful ethnic impact in these lands. Early overseas migration is known from the Classical Greek colonies on the African and Asian littoral and from the

Figure 4.4 Rural settlement types in Europe: (A) street village; (B) green village; (C) irregular cluster village; (D) scattered farmsteads; (E) hamlets; (F) loose irregular villages; (G) linear long-lot villages; (H) gridiron villages

Viking settlement of Iceland and Greenland. The modern phase dates from the Age of Discovery inaugurated in the fifteenth century. Until the nineteenth century, numbers involved were small, but improvement then in transport and communications helped to encourage a wide spectrum of people to move to what were believed to be better opportunities. The main movement was to the Americas – notably by North-west and Central Europeans to North America, whereas Southern Europeans were attracted more to Central and South America (though many also went to the North). Predominantly, British went to Australasia and to the East African highlands, whereas French and Italians tried to establish coherent settlement areas in North Africa and, in the latter case, also in the Ethiopian and Eritrean lands. Although taking place in the same national state, Russian colonisation (much through deportation) of Siberia and later Central Asia should also be included. The total number involved is difficult to calculate, for few official records exist of a largely voluntary movement based on individual decisions, but it is likely to exceed 60 million (almost one-third from the British Isles, with Germany as a close runner-up). The peak of migration across the Atlantic came around 1880–1910, when Russian migration not only to Siberia but also to the New World was mounting rapidly. Between the world wars increasing restrictions were placed on migration to North America, when the United States favoured immigrants from Western and Central Europe at the expense of Southern Europe and the eastern fringe, though within Soviet Russia a rising tide of colonisation to Siberia occurred. There was also vigorous growth of emigration from Britain to the white Dominions. After the Second World War various relief agencies in Europe encouraged emigration, many displaced and homeless people going to North America and to Australasia, with a smaller number to South America, and there was a powerful movement of Jews to Israel. By the 1950s a substantial migration began back to Europe of settlers and officials from newly independent British, French, Italian and Dutch territories. In the 1950s Britain attracted migrants from the West Indies, India and Pakistan, as France drew in workers from Algeria and Tunisia and its former colonial lands; considerable numbers of non-Europeans (notably Turks and Moroccans) went to work in Switzerland and West Germany.

Migration from Europe after 1950 became a more specialised drift of highly qualified technicians, doctors and academics – the 'brain drain' – to the Americas and Australasia and, on a lesser scale, to Africa. The opportunities for emigration from Europe have nevertheless greatly declined as imperial structures have collapsed and even overseas European settlement areas, developing their own clear identity, are more reluctant to admit new settlers.

Overpopulation?

Overpopulation has no fixed thresholds – it is simply exceeding the population that a region can support at a given standard of living in relation to its natural resources and its ability to create wealth. Views in Europe have swung between a fear of overpopulation and a worry about underpopulation. The idea was expressed by Malthus that population might grow beyond the resources to maintain it, but by late in the nineteenth century an ebullient optimism and a feeling of command over nature rejected such gloom. In the interwar years, after the heavy population losses of the 1914–1918 conflict, several states began to worry about having too few people and sought to stimulate growth: for Nazi Germany there was both a fear of too few people and of too little space to live in (*Lebensraum*). Between the wars, however, in many poorer agricultural areas of Eastern Europe, overpopulation (too many people trying to live from the land) existed to which economic diversification might have been the panacea. A feeling of underpopulation remained after the Second World War in Britain and France, where generous policies to stimulate growth were initiated, though in Central Europe (notably defeated Germany) the belief was that overpopulation ruled to which emigration was the only solution.

The feeling of overpopulation – though the basis in fact is uncertain and can be disputed – had begun to emerge once again in the late 1960s, arising mostly from ideas imported from the New World, where a 'doomwatch' approach to world resources and their future had rapidly arisen. The problem is really a complex mixture of elements in which the age–sex structure as well as the numbers of the population are compounded with the ability to generate wealth to maintain an expected quality of life. The situation has often been more structural than real.

Despite limitations of Europe's endowment with many natural resources for an industrial economy, since the Second World War expanding technological horizons have managed to maintain and improve living standards and to achieve this with a modestly expanding population. It is difficult to distinguish whether greater progress has been made in the East compared to the West, since the initial starting bases in 1945 were at such different levels. By the 1970s, however, a question mark came to hang over whether it will be possible to maintain this progress, chiefly because a growing energy availability problem and the emerging bottlenecks in world raw materials supply had become apparent in the West at an earlier date than in the East.

If the economic wellbeing of Europe, or of any part of it, is likely to contract, the appearance of a conceptual 'overpopulation' becomes likely. Overpopulation is an essentially relative concept, but it is invariably recognised whenever living standards in any region or country begin to lag behind the general trend in the larger national or macroregional unit to which the affected area belongs. Consequently, whether in the socialist centrally planned economies of Eastern Europe or the free market economies of Western Europe, policy has increasingly begun to seek to equate economic and amenity levels between regions and in particular to remove inequalities in living standards between town and country, even though the planning philosophies and methods themselves have been markedly different. Overpopulation, expressed as 'job-opportunity', may be seen as the underlying concern of this common planning problem. Certainly, with continuing agglomeration of population and retreating frontiers of settlement in the more marginal lands, it is important to try to control the widening gap between the ends of the wealth spectrum. On the one hand are the wealthy growth regions and their ability to attract people and enrich economic activities, whereas on the other hand are poor regions whose flagging economies fail to provide adequate jobs and economic opportunity and consequently people drift away.

It seems logical to ask if there is any part of Europe 'underpopulated'. Conceptually underpopulation has received little consideration from demographers. For example, any region that can maintain or improve its living standard and yet grow rapidly (whether by migration, natural increase or both) must surely be regarded as still 'underpopulated' – a feature common to several industrial growth areas. But there is also a more serious form – fortunately areally limited and affecting relatively few people – typical of marginal lands, like some small Scottish highland and island communities. Here the drift away of people because of lack of opportunities leaves behind an ageing population declining in numbers and there comes a point when the community becomes too old and too small to support itself or to warrant expensive public services: the decision has then frequently been to remove the remaining people to a larger and more viable community. As agglomeration intensifies and expected levels of public services rise, this type of 'underpopulation' may become more widespread and its thresholds rise.

Towns and Villages

The percentage of population living in towns varies considerably from country to country, though everywhere it is on the increase (particularly rapidly in Eastern Europe). Some countries, such as the United Kingdom and West Germany, appear to be near to saturation level, with well over three-quarters of their populations officially classified as urban. Nevertheless, the distinction between 'urban' and 'rural' has become blurred under pressure from social and economic change.[4]

From late mediaeval times until recently, over much of Europe the village remained little altered, but in many parts recent economic and social changes have triggered big alterations in existing forms. A map of village types reveals a wide range of expression, though basically the contrasts are in the degree of nucleation, particularly as the usual classification is morphological rather than functional or organisational. Nineteenth-century attempts to relate village types to ethnic criteria is an approach now usually rejected. Undoubtedly many enigmas of village origin remain, especially in the process of nucleation or the lack of it. It is clearly impossible to assign one causal factor to any particular village type, but there are likewise few areas where one type alone is found: mixtures with some predominant elements are most usual. Even within particular types a measure of diversity is commonly found.

A generalised pattern of rural settlement shows the contrast between the scattered settlements of upland Britain, related to the long-standing pastoral economy, and the nucleated villages of several types common to lands with a rich arable farming tradition, though in lowlands where mediaeval sheep farming or modern dairy farming have been well developed nucleation is much weaker. There is nevertheless everywhere amid the villages an element of dispersion induced by the effect of extensive enclosure establishing individual farms lying in their own fields. In Scotland, enclosure broke down the small hamlets of farmstead clusters and the villages are relatively recent 'planted' communities (usually the long street pattern as in Strathmore and Buchan).

For Norway and western Sweden (and indeed Iceland) single farms or small hamlets are common and are also found in the Finnish forests. Where villages occur they are rarely large by standards elsewhere in Europe, with the biggest Scandinavian types traditionally in Denmark and Skåne. Nevertheless, in the past two centuries the constituent farmsteads of the villages have tended to disperse, particularly as a result of changes in farming in Denmark, but everywhere land reforms have altered the older patterns. Nucleated villages are still common as fishing communities, usually clustered around their harbours.

Intense study of village settlement has developed an elaborate nomenclature, especially in the Germanic lands, where distinction is made between 'old' and 'new' settlement. The older settlements in West and South Germany are mostly irregular nucleated villages (sing.: *Haufendorf*) and hamlets (particularly tight nucleation occurs where farm land is most valuable, like the wine-growing areas), whereas in the northern plains the eastwards colonisation of the Germans laid out regular villages, mostly variations on the street type (sing.: *Strassendorf, Angerdorf*), which was also used in less orderly form by Slav peoples deep into Russia. The distribution of round villages (sing.: *Rundling*) of an early defensive type in North Germany is related to Slav settlement and its adoption into a German farming economy. A late stage in colonisation was marked by a penetration into the forested uplands, where long linear villages (sing.: *Waldhufendorf*) were laid out along streams, with their holdings stretching back into the forest.

Long linear villages (sing.: *Marschhufendorf*) are a common feature in North-west Germany and in the Netherlands, stretching along dikes enclosing reclaimed land. This was a later stage in colonisation than the small clustered hamlets on artificial mounds above likely flood level (*Terpen, Wurten* or the North Frisian *Halligen*). Dispersed settlement has also spread over much of the eastern Netherlands and North-west Germany (notably Westphalia and Lower Saxony), with either scattered farmsteads or small hamlets on drier glacial mounds amid wet and infertile country.

France's role as a land of transition is seen in village types. In the north and north-west, nucleated but unplanned types similar to West German villages except in house type are common and this form characterises much of Belgium outside western Flanders. Western France (like Flanders) has dispersed settlements with small hamlets, and in Brittany patterns are reminiscent of western upland Britain. Much of southern France is typified by small villages, hamlets and some dispersion, though in the south-west clearing and colonisation in mediaeval times and later is marked by planned settlements of generally geometrical character, whereas the turbulence of events in the south-east is marked by nucleated villages on defensive hilltop sites and a later dispersal into the valleys.

Hamlets and small groups of homesteads characterise the north and north-west of Iberia, whereas large hamlets and small villages are common in Leon and Old Castile, but these grow larger further south, reflecting the safety of

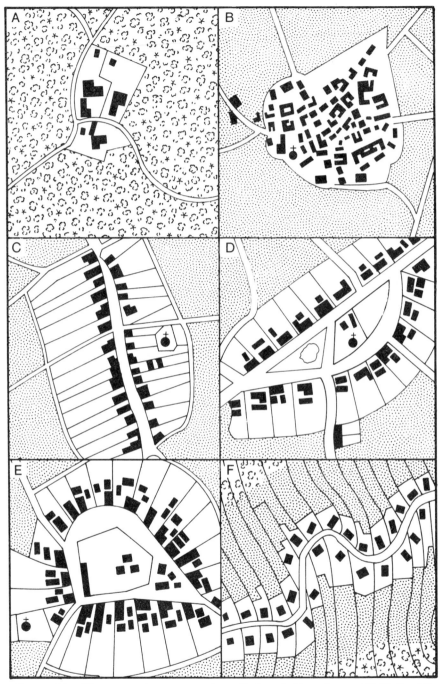

Figure 4.5 Diagrammatic representations of selected village forms: (A) *Weiler*; (B) *Haufendorf*; (C) *Strassendorf*; (D) *Angerdorf*; (E) *Rundling*; (F) *Waldhufendorf*

numbers as the *Reconquista* of the land from the Moors pushed forward. The far south has very large villages widely spaced and Andalucia is marked by large estate villages of *Latifundia*. A wide range of village forms occurs in the Italian Peninsula, but nucleated and defensive forms (often walled) reflect the long disorder of the Italian lands and are particularly common in the Apennines. Generally the south has large villages and even agricultural towns (reflecting the strength of urban influence in Italian life), while patterns in Sicily and Sardinia are similar, though there are also patches of dispersed settlement. The northern plains have a mixed character, reflecting perhaps their role in absorbing different influences through their location and economic importance. But everywhere, the characteristic of Mediterranean village life remains visible in the search for defence and security.

The mountains – the Alps, the Carpathians and the Balkan and Dinaric mountains as well as the Pyrenees – all bear characteristic dispersed settlement, with small hamlets or scattered farmsteads, whereas there are remnants of settlement associated with the dying practice of transhumance. In the Dinaric mountains – but also in the Eastern Alps, for example – long strings of settlements occur, while in all ranges there is often clustering of settlement around and in the broader basins (notably in the distinctive *polja* of the Dinaric mountains). Secluded mountain basins were particularly attractive to settlement in past troubled times. In Greece and along the Adriatic the internecine warfare of the past is reflected, such as in the Italian Peninsula, in clustered villages perched on defendable hillsides or even cliff tops (a feature common in the Aegean islands). In the Bulgarian plains, irregular villages and some more regular forms descended from Turkish *čiflik* estates occur.

The Rumanian plains have loosely nucleated villages of late origin, whereas in many parts of Yugoslavia and elsewhere in the Balkans there is evidence of a shift in the balance between dispersal and nucleation. Nearly everywhere dispersal is a late phenomenon.

The mid-Danubian basin has nucleated villages of a Central European type west of the Danube, with some dispersed settlement, but the long uncertainties of the Turkish period east of the river brought a search for safety in the vast town-like villages, from which in recent times small outlying settlements have been established. In the south, in the Banat and Bačka, land recolonised in the eighteenth century is marked by rigorously planned gridiron villages, whereas in the Sava and Drava plains loose nucleation is common.

This established pattern is beginning to break down throughout Europe. Much of the change comes from new forms of economic organisation – in Eastern Europe, collectivisation has brought much change, with attempts to spread economic diversity and to equate living standards between town and village. New settlements have arisen in an attempt to create the socialist agricultural town, modelled on the Soviet *agrogorod*, especially in East Germany and in Bulgaria. Where peasants have been left more to their own organisation – as in Poland and Yugoslavia – change has been less striking. In Western Europe, many villages have grown into dormitory settlements for nearby towns, forming a new type of suburbia, well seen in southern England and in the Rhine valley in West Germany. In an attempt to put farming on a more rewarding basis to compete with industrial employment, in West Germany

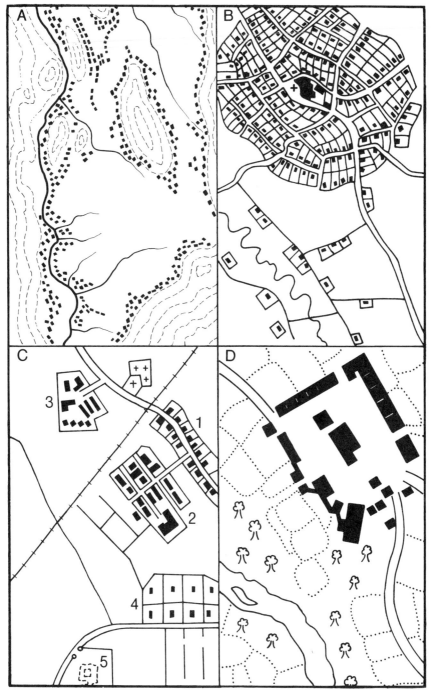

Figure 4.6   Diagrammatic representations of selected village forms: (A) dispersed *polye* (Yugoslavia); (B) *puszta* village (Hungary); (C) *agrogorod*, (1) original village, (2) new housing often as blocks of flats, (3) collective farm buildings, (4) small-holdings created after land reform, (5) original manor house; (D) *ciflik* (southern Yugoslavia)

and in France reorganisation of the village has changed its form. In West Germany, this process has created in many districts self-contained holdings, moving peasants from the village nucleus into new farmsteads on its fringe or even on the edge of the village lands, and in the last quarter of the twentieth century the pace of change has accelerated.

Although not a European 'invention', the town has been carefully nurtured by Europeans to become the most significant element in their settlement pattern. Apparently originating in the Middle East, the town spread into Europe from the eastern Mediterranean, first through Classical Greece and its colonies and subsequently through the towns widely established by the Romans, from whom the urban concept was kept alive, and ultimately in mediaeval times spread much more widely. The dissemination of towns to all parts of Europe from the Eastern Mediterranean lasted well over a millennium, occurring latest where social and economic conditions were least attractive. Regionally and over time, the town has taken many different characters and no real understanding of Europe is possible without some knowledge of the historical evolution of its towns.

The proportion of urban population (table 4.2) is highest in Western Europe and declines towards the south-east. Such percentages are, however, only a guide, since the statistical concept of urban population varies from country to country, just as the legal definitions alter. The rate of urbanisation is also different – in Western Europe, where in some areas over 80 per cent of the people are urban, the share of total population in towns is rising only slowly, for the massive explosion of urbanisation in the last century has died away. The question remains whether one can expect countries or major regions to become completely urbanised in a truly geographical rather than merely statistical sense. The drift to towns is still strong in Eastern and South-eastern Europe, where urbanisation is continuing at an appreciable pace. Even where the proportion of town dwellers is still modest, such as South-eastern Europe compared with Western Europe, this proportion is

Table 4.2   *Percentage of urban population in European countries (according to latest statistical information)*

Belgium	87.1	Finland	57.7
Sweden	81.4	Northern Ireland	54.7
F.G.R.	79.3	Switzerland	54.6
England and Wales	77.7	Poland	54.4
Netherlands	77.2	Ireland	52.2
G.D.R.	75.4	Austria	51.9
France	70.0	Hungary	49.7
Scotland	70.0	Norway	44.8
Luxembourg	68.9	Rumania	42.7
Denmark	66.9	Yugoslavia	38.6
Czechoslovakia	66.7	Albania	33.8
Greece	64.8	Portugal	26.4
Bulgaria	58.7	Spain	n. a.

nevertheless generally higher than in Africa and Asia.

Europe is distinguished by its richness in towns, both in numbers and by their variety. Many older towns still display clearly their several stages of growth and there is a rich diversity of functions and site elements. The first towns in considerable numbers were founded by the Classical Greeks and copied by others, and were usually small, often hardly distinguishable from villages, with the difference often more cultural than economic. The real spread of towns was the work of Roman civilisation, whose imperial structure was based upon towns as the key points in its political geography and as disseminators of Roman ideas. Towns were the administrative and military foci in conquered territories. Most were neither large nor impressive, but they had regular street plans and communal facilities such as baths that were not found commonly again until the eighteenth century. Although many Roman towns were to perish (and in the Balkans even their sites have been lost), a considerable proportion has provided, at least, sites for later towns, perhaps reflecting the remarkable Roman eye for terrain. Debate continues around the extent of continuity of use – did Roman towns simply metamorphose into Frankish, Saxon or other peoples' concept of the town or were they for periods completely deserted? Possibly the whole spectrum of continuity occurred. Certainly in Southern Europe, Gaul and even the Rhineland, modern towns still display elements of their Roman predecessors, while Roman walls provide the base of many mediaeval fortifications, even in Britain.

In Dark Age Europe, amid the rotting Roman towns, it was the Church or the seats of princely power that kept the urban tradition alive, closely bound with the defensive role. To these foci artisans, merchants and markets were attracted: only the wealthy princes and their households, secular or ecclesiastical, could afford or needed their wares in contrast to the subsistence economy of the countryside. Nevertheless, urban life survived – best in Southern Europe (in Italy, Iberia and Southern France), but despite immense difficulties and destruction also in the Frankish lands, where a new virility was imparted that ultimately carried the town through German colonisation and influence into Eastern Europe and beyond well into Russia, while the Germans also played a significant role in taking the town into Danubia. Many archaeologists, especially in Eastern Europe, would now claim that large settlements with some apparent urban functions and characteristics did, however, exist among the Slavs before the influence of the Germans spread.

Although they varied from region to region in detail and in appearance, Europe's mediaeval towns had several recurrent features in common. An early focus of agglomeration was a stronghold that attracted to it artisans and traders, but at some sites it was the Church and its appendages, where the 'immunity' afforded by an ecclesiastical precinct often made nearness to it an asset. Artisans and traders usually managed to generate a corporate organisation that in the pattern of feudalism brought them recognition and a charter as a town: others not accorded rights and membership were forced to live extra-murally until their own identity could be established in another walled community. In some towns, three or more separately chartered quarters may be defined. Interaction of a wealthy merchant class and a well-organised Church was not uncommonly reflected in splendid ecclesiastical buildings –

cathedrals, monasteries and bishops' palaces.

Of all the factors attracting agglomeration of human beings into town organisms, most powerful was exchange – the interface of the market. Trade was the ultimate hallmark of urban success, even in towns with important administrative or other functions. The market place usually contains the town hall (the sign of emancipation of the burgesses from feudal ties that bound lesser orders), the wealthy patrician houses, the warehouses and, nearby, the workshops. The most successful trading towns usually developed more than one market place through the division of specialisation of trade. Powerful territorial lords encouraged the foundation of towns to attract trade and consequently wealth to their domain – the mediaeval period is rich in such planned and planted towns. In North Germany, for example, it seems that many market stances and sites of fairs were given charters even before a proper urban community had developed. Town foundation in South Germany was so prolific that many today remain as dwarf towns, ossified mediaeval communities little bigger than villages (for example, Schlüsselfeld am Main). In the layout of towns, particularly in the planned towns of the German 'colonial' lands east of the Elbe, effort was made to ease the flow of goods from the river quays to the markets and the warehouses by careful design of the street pattern (for example, Elbing – now Elbląg). Towns became an important mediaeval technique of developing territory and stimulating economic development (often behind a defensive role) – the 'bastides' of North Wales (Conway) or southern France (Aigues Mortes) and the Trans-Elbian 'colonial' towns (Neubrandenburg) are good examples, while Ireland abounds in 'planted' towns.

Bases of trade, centres of defence, focal points of spiritual life, it is hardly surprising that towns became the cores of administrative power and seats of justice. Learning was kept alive by the Church and universities began to appear from the mid-twelfth century, giving selected towns a special and distinctive pattern. In early mediaeval times, royal administrations in Western Europe were peripatetic, moving between selected royal estates (the centres of some were later to become towns), but usually one town came to be more favoured than the others (often for no easily apparent reason) as a permanent seat. From this arose the 'great capitals', sometimes the result of the success of one faction over another in an emerging national pattern.

Towards the sixteenth century the wave of town foundation had largely passed, but new ideas were developing. The chaotic streets, narrow and twisting, darkened by overhanging buildings, the result of an insufficiently strong town government interested solely in material gain, began to go out of fashion. A rediscovery of Classical learning began a search for order and symmetry. The market place remained important, though economic and financial mechanisms had become more sophisticated; while the Church still demonstrated its spiritual power through displaying its material wealth. But gunpowder and cannon had rendered the old walls and castles ineffectual – new ideas of urban defence emerged, very much through the work of the French engineer Vauban. Now there were created vast systems of ramparts, glacis and bastions to allow enfilade of attackers. In the great fortress towns, the low profile of the defence works not uncommonly exceeded in area that of the

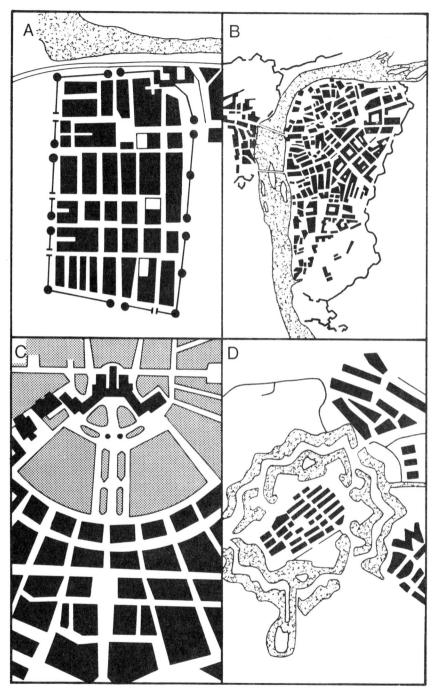

Figure 4.7 European town types: (A) Aigues-Mortes (France); (B) Prague (Czecho-slovakia); (C) Karlsruhe (F.G.R.); (D) Naarden (Netherlands);

town and its barracks. The withering of feudalism resulted in an acceptance of the town into the structure of the state as the basis of its power: rulers lavished immense sums on building towns (for example, Karlsruhe, Paris, Berlin, Warsaw) which in their eyes suitably demonstrated their wealth and omnipotence, while lesser princes drained their subjects dry to emulate them (for example, the small German *Residenzstädte*, such as Oldenburg or Schwerin).

Well into the eighteenth century, towns over most of Europe seldom claimed more than 20 per cent of total population and even their number was modest. The great economic revolution of the next century was to change this picture radically over large areas, beginning in Britain and spreading into Western Europe and then eastwards, as factory industry and the consequent widening of trading and commercial horizons developed. Not only did existing towns in most instances show rapid growth, but there was also a rash of new towns, many arising from charters issued to existing settlements, and town life became rewarding for the masses in a way never previously experienced. Growth of population in towns was much quicker than overall national growth and a vast stream of migrants from country to town was generated, often over considerable distances such as the movement from East Germany to the Ruhr towns in the 1875–1914 period. The industrial towns, a loose aggregation of mills and works, with their workers' housing, split by railways and canals, often had little definite centre and were sometimes more an administrative convenience than a truly defined community (examples abound in industrial England and in the Ruhr and Silesian coalfields), bringing together small industrial settlements into a single 'charter'. The smaller industrial towns, including nearly all the greenfield foundations, were generally centred on one industry, even one plant, and sometimes even owned by one company, which had often created them (such as New Lanark, Middlesbrough, Rjukan, Kiruna).

More diversified towns – with industry and one or more other activities – showed generally well-defined centres, though the market place gave way to more sophisticated forms of commerce that produced an identifiable 'central business district' of shops, banks and offices. Some towns became associated closely with specific commercial operations; for example, Manchester as the organising and commercial centre of the cotton textile industry or Leipzig with printing and publishing. As many towns grew far beyond the population size of even the greatest mediaeval cities, subdivision and specialisation in their commercial districts became commonplace, but one of the most impressive developments in the twentieth century has been the growth of the tertiary sector and in particular of administrative functions. Noteworthy has been the position capital cities have come to occupy in most countries, where they are usually the 'primate city', claiming a considerable part of the total national population (for example, Austria 22 per cent, Hungary 20 per cent, Denmark 28 per cent).

Two elements traditionally important in the town have weakened – the Church and defence. Late in the nineteenth century elaborate annular defence works were scrapped and their lines made into wide avenues and streets (for example, in Cologne and Gothenburg), though they were occasionally replaced for a short period by a ring of forts (themselves razed within the first quarter of this century). The great churches remain and ecclesiastical quarters are found in many older towns still, but few major new churches have been built (Liverpool is perhaps an exception, while the cathedral in Cologne was only completed in the later nineteenth century). On the other hand, educational facilities have continued to grow with the establishment of many new universities and higher technical schools, especially in a major boom that came in the 1960s. Nevertheless the European towns, compared with those of North America or Australasia, are extremely compact and economical in their use of space, though in the mid-twentieth century a more generous use of space to give modest residential densities at the expense of rural environments has developed.

As we see, through the ages the concept of the town – the search by town councils to create their ideal – has changed, while even between the different parts of Europe at any one time the town has been viewed differently in conceptual terms. A new interest in the idea of the ideal urban environment arose, however, from the squalor of many nineteenth-century industrial towns. Associated with the idea of the 'garden city' are particularly the names of Fritsch in Germany and Ebenezer Howard in Britain, the products being, for example, Welwyn Garden City and Letchworth or Bournville in Britain and Krupp's Margarethenhöhe or Siemensstadt in Germany.

Ideology has also played a part in town design: Nazi concepts that sought to bring the virtues of the countryside into an urban setting were seen in the never completed plans for the new industrial towns of Wolfsburg and Salzgitter or even the grandiose Speer Plan for prewar Berlin. Since 1945 there has developed the concept of the 'town of socialist realism' in Eastern Europe, based mostly on Soviet ideas. As the ideological vanguard, it must be 'socialist in content, national in form', and design is strongly based on the neighbourhood unit (to reduce wasteful intra-urban movement) and focuses upon the main industrial plants, though living space is usually separated from them by a green belt. The central business and shopping district so important in Western European towns is minimised in the centrally planned economy, so reducing the number of shops (now provided on a strict ratio to population) and service outlets such as banks and insurance offices, and yet there is a strange emphasis on 'the big city effect' in the design and spacing of buildings and thoroughfares. As examples we may cite Dunaujváros (Hungary), Eisenhüttenstadt (G.D.R.), Nowa Huta (Poland), Victoria (Rumania) or Kremikovci (Bulgaria), though the elements have emerged in other towns through postwar redevelopment. These new towns of *Comecon* form a major contrast in concept to the many new towns of Britain (15 created in the first postwar decade and others later), Sweden, France and West Germany (though a relatively rare phenomenon, like Hochdahl near Düsseldorf).

Late twentieth-century Europe may well be called the 'urban continent' – not only in terms of the proportion of people living in towns, but also through the extent of the built-up area and the rapidity of its spread, the pace of which has accelerated. One of the most impressive features is the spread of agglomerations of towns and suburban sprawl into homogeneous urbanised landscapes, in many cases bordering on 'megalopolis'. In the German Federal Republic, planners have expressed real fears that the town is disintegrating

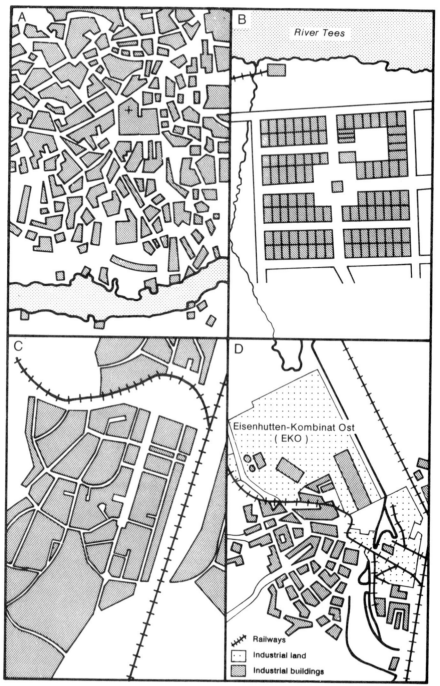

Figure 4.8   European town types: (A) Toledo (Spain); (B) Middlesbrough 1830 (U.K.); (C) Welwyn Garden City (U.K.); (D) Eisenhüttenstadt (G.D.R.)

into the countryside; Britain has sought to contain towns by rigorous controls to prevent the featureless sprawl experienced in the 1930s; whereas the French have tried to contain town growth in selected regional centres to infuse new life into *le désert français*. Planners in East and West see the town as the key point in the settlement pattern, and towns by their nature and function allow themselves to be grouped into nesting hierarchies of communities. Towns have mostly become the growth points in social and economic terms, whereas the countryside between them is increasingly grouped into problematic development areas. Planners now seek to link towns by growth axes supposed to bring new life to the countryside through the spin-off from their pulsating activity. The growth point–axial concept was first classified for South Germany by W. Christaller, where it is particularly appropriate in the long stable settlement pattern, though other parts (for example, North-east Scotland) appear to possess modified and less perfect hierarchies of the type, but it becomes confusing and complex to apply it to many industrialised landscapes such as the English Midlands, the Lower Rhine or even the Plains of Lombardy. Nevertheless, towns and the axes between them, however complex in structure, form the framework of circulation of people and goods as well as ideas within Europe.

References

1. Detailed examination of these aspects is to be found in D. Kirk, *Europe's Population in the Interwar Years*, League of Nations, Geneva (1946), and L. Kosiński, *The Population of Europe*, Longmans, London (1970).
2. L. Kosiński, *The Population of Europe*, Longmans, London (1970).
3. Early stages in population and settlement development are reviewed in C. T. Smith, *An Historical Geography of Europe before 1800*, Longmans, London (1967).
4. Settlement is examined in detail in J. M. Houston, *A Social Geography of Europe*, Duckworth, London (1953); R. E. Dickinson, *The West European City: A Geographical Interpretation*, 2nd edn, Routledge and Kegan Paul, London (1964); and L. Mumford, *The City in History*, Secker and Warburg, London (1961).

5 The Economy to 1939

The evolution of the present economic geography of Europe has depended strongly on local natural resources and the ability to use these to provide goods and services to sell in the market in order to buy those raw materials, foodstuff, goods and even services locally deficient or absent: this applies at all scales, from the village to the town, from the province to the state, from the local dimension to the continental dimension. The uneven and disparate regional distribution of local resources has much affected the pattern of economic geography, especially during the massive wave of industrialisation in Western Europe and the corresponding somewhat later development in Eastern Europe.

The Raw Materials Base

Energy is a key to economic development. Until the coming of the steam engine, water power turned mill wheels to grind corn or even activate the furnace bellows. Some modest rivers of the plains, whose flow was not too powerful to control, were harnessed by building weirs or, as on the Danube, setting waterwheels in boats where the current was reasonable and driving mills on the banks through long shafts. Wherever wind was reliable, windmills were built – often with great skill to make the best of the feeblest breeze – in North-west Europe, in Eastern England, or on the Mediterranean islands under the Etesian winds. The invention and application of the steam engine in the eighteenth century was, however, the key that made the massive nineteenth-century industrialisation possible: the most suitable fuel was coal, whose distribution came to dominate industrial location until the mid-twentieth century, so that the map of major industrial areas in the golden year of 1913 shows a remarkably close association to the coalfields. The wide scatter of industry in Britain was closely related to the many coalfields – over eighteen in all – whereas the smaller number of individual coalfields within the German Reich was reflected in the fewer clearly industrialised districts; while in France, industry gathered on the major coal resources which lay predominantly in a strategically dangerous northerly frontier position. Few large coalfields were known to exist in Eastern Europe and one of the most significant in Upper Silesia lay across the boundaries of the German, the Austrian and Russian empires, drawn before the real implications of coal were appreciated. It was only in the mid-nineteenth century that large coal deposits in Russia were to be opened, in whose eastern territories deposits are still being mapped. The decline of the role of coal in energy generation that set in during the 1930s and rapidly accelerated in the 1950s arose far more from economic reasons than from any exhaustion of resources (for large unexploited reserves still remain in Britain, West Germany and several other countries) and may be reversed in the future.

In Central Europe, large reserves of brown coal (lignite), a fuel of much lower calorific value than bituminous coal, are known and have been exploited on a major scale since the latter nineteenth century. The rise of industry on the Lower Rhine, in Saxony and in parts of Bohemia and western Poland was linked to this raw material for energy. Because it was unsuited to metallurgical use, the brown coalfields have been associated primarily with the generation of electrical energy and with the chemicals industry.

Generation of electric current by thermal means grew rapidly in the last 20 years of the nineteenth century and immediately before the First World War, with the main concentration of generating plant on the coalfields, though generators fed by rail-borne coal were built in most major towns. Use of water power by turbines to generate electricity favoured countries with large harnessable supplies and several, notably the Scandinavian countries, Austria and Switzerland, thus found a solution to their inadequate endowment with coal. The mountain areas in particular, with fast-flowing but not overpowerful streams and valleys suitable for housing reservoirs, became associated with hydroelectricity, though in Scotland such development came only in the 1940s.

The late nineteenth-century development of oil wells in North America and in southern Russia also brought a search, though of no great urgency, in Europe. Discoveries with the simple technology available were small by modern standards. Before 1939 the major petroleum producer was Rumania, but some came from southern Poland, parts of the North German Plain and Upper Rhine valley, as well as the mid-Danube basin, and Russia was also again beginning to emerge as a major world producer (a position lost during the First World War and the Revolution) as new deposits of great size were mapped in the Volga basin. The low level of domestic demand could usually be reasonably covered in the East Central European countries from their own small resources, whereas countries with serious shortfalls (such as Germany and the United Kingdom) had no great difficulty importing refined products from refineries overseas, often from their own colonial territories or where they had marked economic or political influence. A growing strategic consideration in ill-provided countries, especially Germany, in the 1930s led to development of synthetic petroleum production. It would be untrue, however, to suggest that a critical 'energy crisis', even piecemeal, existed anywhere in Europe before 1939.

European endowment with metallic minerals, particularly for modern industry, is far less attractive and despite a wide variety few are available in quantities to satisfy demands, and most occur in high cost deposits. Since the late eighteenth century, Europe has become increasingly dependent on outside sources of metallic ores and raw metals. Despite gold and silver being found, it was the rising demand stemming from mediaeval prosperity that drove Europeans (especially the Spanish, Portuguese, Venetians and Genoese) to search for sources beyond the sea, even though rich mines had been worked in the Central European uplands and in the Carpathians. Iron, one of the most commonly used metals, is widely scattered in deposits of several types and was reasonably abundant throughout mediaeval times, but limitations to smelting were commonly posed by availability of charcoal fuel. In this way, iron making declined from the fifteenth century through exhaustion of the Central and Western European forests, and iron began to be imported from Sweden and Russia. Some ore deposits were unworkable because they contained much phosphorus that made brittle iron; consequently, many sedimentary deposits were of little interest until appropriate techniques were developed in the nineteenth century.

The fuel problem for smelting iron was a key factor in the search for a way to use coal, first successfully applied in England at Coalbrookdale, and this greatly widened the number of locations where iron might be smelted. The use of coal was even more attractive where it and blackband iron ore could be worked in the same mine, as in the English Midlands and in the German Ruhr. Otherwise a close association was sought: the Ruhr coalfield smelted ores from the nearby deposits in the Siegerland and Lahn–Dill valleys to the south; the Durham coalfield used the Cleveland ores; while Czech Silesia used ores from nearby sources in Bohemia and Slovakia. The closeness, by Russian standards, of Donbass coal and Krivoy Rog iron ore brought rapid growth in the 1880s of a modern iron and steel industry that quickly overshadowed the traditional charcoal industry of the Ural, despite its renowned quality. By the end of the century, similar relationships were leading to the growth of iron smelting in northern Hungary, in Bosnia and Slovenia and even in Rumania. When it became possible to use phosphoric ores after 1879, new orefields were opened, generally not too distant from coal, as in Luxemburg, in Lorraine, and in the eastern scarplands of England: it was ores of this kind that in the 1930s encouraged the Germans to develop the large new steel town of Salzgitter. The success of the iron and steel industry in producing quality goods at an acceptable price so stimulated demand that, by the second half of the nineteenth century, ore was beginning to flow over considerable distances if its quality was high enough. Spanish ores were carried to Britain and to Germany, which in addition took Swedish ore that also became an important supply for Central Europe (for example, to Silesia).

The older uplands and shields have long supplied metals other than iron – notably copper, tin, lead and zinc, and even precious metals. From ancient times, the lead mines of the Pennines, the tin mines of Cornwall, the copper mines of Central Germany as well as the Spanish mines had been important, even if quantities were small and demand modest. Lead mining had been particularly important to the Romans, voracious consumers for coffins, water pipes and cisterns. In mediaeval Europe, demand for metal encouraged the migration of miners (such as the Germans invited into Bohemia and the Carpathians by Slav princes), but in many parts, especially in the Turkish Empire, methods were backward. The rich potential of the Balkans and the Carpathians was only to begin effective development in the nineteenth century through the infusion of capital from Western Europe, as for example in the copper mines of Serbia. As these resources had been barred to Western Europe by the Turkish hold, the search for non-ferrous and precious metals had been a powerful driving force in European exploration overseas from the sixteenth century. Some metals depended, however, on appropriate technology before they could be effectively used, as for instance the bauxite deposits of southern France or Hungary in the growth of aluminium, or even the extraction of metals as by-products of other mining, such as cadmium.

A major nineteenth-century industrial development in Europe was the factory production of a wide range of chemical substances using the continent's natural resource endowment. Salt mining, one of the first incentives to commerce, took on a new significance and began to include many mineral salts not previously used but available in rich deposits. By the mid-nineteenth century, some raw materials like pyrites could no longer be found in sufficient quantity to satisfy demand and imports began. The latter part of the century saw advances in the use of coal and lignite for chemicals production, especially for tars, phenols and dyestuffs (of which Germany by 1914 had a near monopoly). Using phosphatic minerals, artificial fertilisers were increasingly important, especially as supplies of natural guano became difficult to obtain, and various nitrogenous chemicals began to be made synthetically. The locational factors of this industry were, however, not just nearness to raw materials (often carried distances by rail and water), but nearness to good process water and energy supplies. By 1914, it was largely concentrated, however, in Central Germany and the Rhine valley, or in the North-west, North-east and Humberside of England, though it was growing in Belgium and in northern France.

Development of Trade and Industry

Until the innovation by Europeans of large-scale industry on a factory basis, small artisan crafts predominated, often with a strong domestic bias, though products were sometimes highly specialised.[1] In Classical times, items of specialised local production (such as pottery) were traded widely in the Mediterranean basin and this remained true in the Middle Ages as well, with quality and market control exercised by elaborate guild systems. Though most everyday requirements were made locally, a growing horizon of trade shifted from luxury goods to more everyday items over the period. The revival of economic life after the Dark Ages showed first a marked focus in Northern Italy, where a vigorous merchant class encouraged manufacture through investment and sales – Genoa and Venice, the great trading towns, as well as Milan and Florence were the most notable centres, trading in the Mediterranean basin and with such great fairs as those of the French Champagne country and with South German towns. Florence was noted for its cloth, Milan for its weapons, and Venice had a highly developed shipbuilding industry, though centres elsewhere, like Toledo and its steel makers in Spain, were also renowned. The total volume was, however, small: even at its peak, Venice handled little more than a few hundred tons of goods annually.

The growing problems of trading in the Turkish sphere in the fifteenth century and the wider horizons offered by the great discoveries in the Atlantic – in the Americas and in Africa – in the sixteenth century brought a shift in interest and a decline in the fortunes of many Mediterranean producers. The growing importance of the Atlantic coastal areas, particularly Flanders and England, as well as the widening trade with Eastern Europe (from which raw materials were beginning to be drawn), brought a period of prosperity to the South German towns such as Augsburg, Ulm, Nuremberg, Nördlingen, Rothenburg and Swiss St Gallen, all well placed to join in the still-wealthy Italian trade, often by using Italian artisans or the sophisticated financial techniques developed in Italy. The extensive woollen industry of Flanders had been established by the twelfth century and survived through its favourable trade location, able to draw on the wool supplies of North-west Europe, even though the manufacture of woollen cloths spread as Flemish weavers migrated to exploit new markets. During the thirteenth and fifteenth centuries there was a concentration in the major towns, notably Ypres, Ghent, Bruges and Douai. The increasing affluence of North-west Europe tended to widen the market among artisans and peasants, raising demand for middle-quality goods in ever greater quantitites rather than expensive qualities on which the Italians had concentrated. As these new demands were for quantities of bulky goods, transport from producer to consumer became a more important consideration, drawing the two together. Rising demand for metal (including silver for currency) encouraged mining and brought metal-working nearer to the mines, as in Central Germany and Bohemia, or the later growth in the Russian Ural and in Sweden.

The perceptible shift of manufacturing to Northern Europe in the fifteenth and sixteenth centuries was a combination of innovation and the absence of inflexible guilds that made new developments easier, while later Protestantism, with its absence of hostility to change and veneration of hard work, was probably a further force in change. Many Protestant artisans persecuted in Roman Catholic territories sought refuge among people of like conviction, carrying with them their skills, as in Prussia. The new manufacturing of the north also aimed at the increasing potential of the middle and lower class market. The initiative in trade had also shifted to northern waters, to the English, Dutch and for some 150 years to the *Hansa*, which in its heyday had over 100 members scattered across Northern Europe. The trade in salt fish from the Baltic and North Sea to southern Roman Catholic countries was particularly significant.

From the late sixteenth century to the eighteenth century, social disruption and wars petrified large tracts of west Central Europe, stagnating trade and hampering industry. The great mediaeval trading axis from the Low Countries to Northern Italy across eastern France, the Rhineland and the Alps was replaced in importance by seaborne trade, to the benefit of the Netherlands and England. Baltic trade also rose (its peak came in the mid-eighteenth century), supplying copper and iron as well as tars, resins, waxes and wood, all in demand by the new industries.

The Age of Discoveries had opened up entirely new possibilities. The Portuguese developed new trading links with Africa and the East Indies, while the Spaniards began trading with the Americas, in which the Portuguese also had a share. Later, especially in North America and in the West Indies, but rapidly spreading to other spheres despite Iberian opposition, the British and Dutch began to trade and were quickly joined by the French.

The new importance of maritime trade had shifted interest to ports and coastlands around the Atlantic west, though inland trade remained vigorous. Difficulties in the Mediterranean, however, saw trade there decline and with it the fortunes of Venice and Genoa. New maritime routes replaced the old caravan ways to the East (through which the Black Death had come) and even the nineteenth-century opening of the Suez Canal was to do little to restore the fortunes of Mediterranean trade, only Marseilles and Genoa retaining rank as first-class ports. The seventeenth and eighteenth centuries were distinguished by an economic nationalism emerging among countries with overseas possessions, which sought to contain trade between themselves and their colonies, exchanging European manufactures for colonial metals or subtropical or tropical plantation products. It was this mercantilist system that brought the ultimate rift between Britain and the North American colonies that did not fit well into such a concept.

Seventeenth-century economic conditions tipped the balance of costs and prices increasingly in favour of industry, while markets began to change from local town spheres to a national scale. With a growing spectrum of trade, regional specialisation became more common, so that organisation became more complex and investment grew. Entrepreneurs began to supply craftsmen with raw materials and to market their produce, a system that encouraged use of more than locally available raw materials. In branches such as mining, shipbuilding and smelting or brewing, numbers of workmen came to be employed together, laying the foundations for the factory system, which was also encouraged by technological progress that required larger investment. States encouraged such organisation to maintain quality or retain monopoly. Even before the revolution brought by the steam engine, large units were already found in England and even in France (for example, cloth making at Abbeville, mining at Anzin), while Prussia was doing the same in its 'arsenal' of Silesia. Migration played a role in dissemination of industry: English entrepreneurs and skilled workers began to develop industries in France, the Low Countries and even in Prussia and Russia (which had always relied on foreign skills).

In mediaeval times, craftsmen and artisans had gathered in the towns, but in the seventeenth and eighteenth centuries, industry was scattered between town and country until the factory system in the nineteenth century again forced concentration in towns, though many industrial villages themselves grew into towns. This development was common in the textile districts of Lancashire and Yorkshire, as well as in Saxony and Silesia, while a similar pattern marked some metal-working areas, such as the English West Midlands or the German districts of Berg and Mark, and in France comparisons could be made with the northern textile districts or the metal-working of St Étienne. This pattern was also seen in the Swedish iron industry of Bergslagen and the glassmaking of Bohemia.

The remarkable outburst of innovation and entrepreneurial genius that began in eighteenth-century Britain set the pace for industrial development that spread eastwards through the nineteenth century, using constantly more and improved machinery and the business concepts of the 'industrial revolution'. A major development was the steam engine, which provided a much greater potential of power and energy than traditional water power. From the 1790s steam engines spread quickly in Britain and were soon taken up on the Continent. They enabled mining to go to much greater depths and become more productive by better pumping and hauling gear, and by driving machinery confirmed the factory system, giving it an immense advantage over domestic industry. Steam was to provide a revolutionary application to transport (an idea grasped in Britain and quickly thereafter in the United States before its full impact was realised on the Continent). Coal for steam engines was the focal point of factory industry for over 150 years.

As reflected in table 5.1, Britain had few challengers initially, but there was undoubtedly the factor that progress was retarded on the continental mainland through its isolation by the Napoleonic wars and the 'continental system', despite attempts to the contrary. Once stability returned after 1815, progress was rapid, especially as Germany found economic unity. The new Prussian territories in the Rhineland and Westphalia, with their coal and iron ore, not only had local ironmasters and entrepreneurs of vision, but also the keenness of the Prussian state to develop industry, as it had already done in Silesia and was to do in the Saar.

The early nineteenth century saw the emergence in Britain of the free trade concept that encouraged the fall of barriers to trade and fitted well with the British position of strength as the 'workshop of the world', though most mainland continental countries, fearing British competition and economic supremacy, were becoming more protectionist. Nevertheless, the industrial and agricultural revolutions everywhere fostered trade – raw materials for industry and food for the rising urban and non-agricultural population could be exchanged for manufactured goods from the new factories.

Effective industrial development on the continental mainland got under way in the 1860s. The strongest upsurge was seen in Germany, the outcome of new economic unity and Prussian drive, with vigorous developments in the Rhineland, in Saxony and in Silesia, whereas an added impetus came from the successful wars against Austria in 1866 and against France in 1870. French industrial development, partly from the military disaster of 1870, faltered. The Habsburgs encouraged industrial development, though not all their territories were socially and financially in a position to pursue it. Coal and iron ore, entrepreneurial skill and a tradition of domestic industry, quickly made Bohemia and Austrian Silesia the focal point, whereas after 1867, Hungary, seeking equal status with Austria, developed its own limited coal and iron ore resources (mostly in the north, in Slovakia and in Transylvania). Russia depended much on foreign and Jewish initiative, with textile factories growing in Russian Poland, where the Dąbrowa district became a main centre for iron, steel and engineering products; but there was substantial Russian and foreign investment in St Petersburg, Moscow and district, and in the Baltic lands.

Europe in its Golden Age of Industry

Although, in 1914, it was clearly Britain and Germany that dominated the European economy, industrial development was growing apace in many

Table 5.1. Changing economic potential of Europe's major powers 1840–1914

Country	Coal output (million tons)				Pig iron (thousand tons)				Population (millions)		
	1840	1855	1870	1914	1840	1855	1870	1914	1840	1870	1914
Great Britain	32	65	117	292	1 420	3 000	6 059	10 400	16.1	31.3	45.2
Germany	4	8	33	190	143	325	1 391	17 600	24.2	40.3	68.0
France	3	7	16	41	348	600	1 178	5 200	26.8	36.0	41.0
Russia	–	–	9	29	–	250	300	4 200	38	75	157

Sources: various.

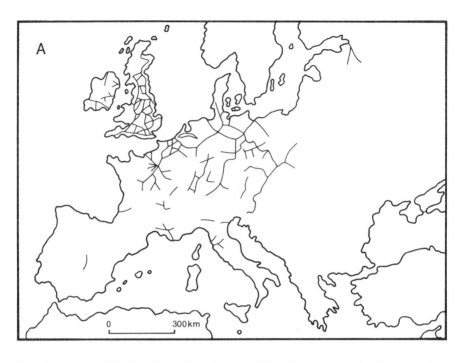

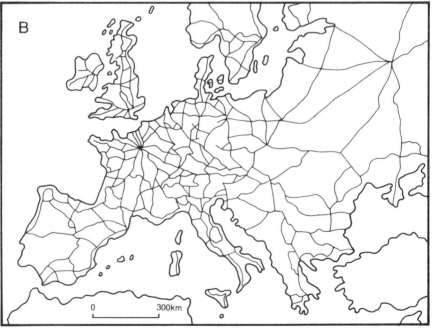

Figure 5.1 European railway network: (A) *c.* 1850; (B) 1978

parts. It is useful to attempt to portray the main industrial districts in that critical year, the climax of the Age of Coal and Iron. Most of the British industrial concentrations were identifiable with a coalfield, but nearly all were close to seawater. Scotland's Clyde basin iron and steel industry, shipbuilding and engineering were thriving; North-east England was a major coal exporter as well as making iron and steel and chemicals, while it also had a successful shipbuilding industry. North-west England was best noted for its Lancashire cotton industry, but there was also coalmining, chemicals (using Cheshire salt and Derbyshire limestone), iron and steel, shipbuilding and engineering, and a small outlier of heavy industry lay in Furness and the Cumbrian coast. Yorkshire had its well-known woollen industry, but there was also iron and steel making (with the Sheffield cutlers' tradition), coalmining and varied industries (including chemicals) around the Humber. The English Midlands had scattered coalfields with the remains of an early iron-making industry, but also engineering and pottery, while South-west England still had its old textile industries. South Wales, a major coal exporter, had an iron and steel industry with a reputation for tinplating. South-east England was beginning to show new industries in an emergent consumer society.

In Germany there was a clear line of industrial concentrations stretched east–west across the *Reich*. Around Aachen, there was coalmining, textiles and some engineering, but the Ruhr coalfield was the main heavy industrial concentration – with its rapidly growing iron and steel and engineering and chemicals industries. To the east, the Hannover district was expanding, with engineering and iron and steel making. A significant concentration with rapid growth features in the chemicals industry was found in Saxony, where engineering and textiles were important, with brown coal as the energy base and a rapidly rising output of electric current, influenced by the technical developments of electrical engineering plants in Berlin. Still further east, Silesia had its textile industries, its engineering, its iron and steel and its coalmines. Emergent nodes lay along the Rhine, mostly with manufacturing industries, but there were grandiose plans for middle-Rhine iron and steel plants using Ruhr coal and Lorraine ore. The Saar was a well-established iron and steel area, thanks to Prussian efforts. The Germans, in the 40 years since 1870, had done much to develop Lorraine and Alsace industrially.

France lay well behind in the industrial power game. Apart from the northern coalfield, a secondary producer compared with giants like the Ruhr and the Eastern Pennines, there were no other notable coalfields in France. Although textiles, engineering and some iron and steel making were found in the north, much industry lay further inland, especially the strategically important Le Creusot and St Étienne works, while Lyon was a major textile centre. There was considerable industry in the main ports, but plans formulated to develop iron and steel making on a considerable scale in Normandy never materialised. Engineering, shipbuilding and textiles were well represented and scattered in Belgium and Holland. With better endowment with coal and easier access to iron ore, Belgium had the more significant iron and steel industry, including the Cockerill works at Seraing, established in 1817, one of the first of their kind on the continental mainland.

By 1914, industry was making great strides in Russia, though Russian

Poland was the main industrial base, with textiles in Łódź and in Bialystok, engineering in Warsaw and iron and steel making in the south-west around Dąbrowa. Moscow had scattered around it a considerable textile industry, but the largest mill was at Narva. St Petersburg, Moscow and Riga had engineering plants (mostly German financed) and several provincial towns had good-sized railway works. By 1914, over three-quarters of total iron and steel production came from the Donbass coalfield, joined since 1884 to the Krivoy Rog iron ore by railway. Plans were afoot to link the iron ore of the Ural to the coal of Western Siberia, but they were not to be realised until the 1930s. Russia was still one of the world's major oil producers, obtaining its oil from the wells of Transcaucasia and Groznyy.

In 1914, coal and iron production in the Habsburg Empire lagged well behind Russia. Iron and steel making was chiefly in Czech Silesia, central Bohemia and in the northern Hungarian plants – all rather small. Slovenia had some small ironworks and development had begun in Transylvania. Engineering was mostly in Vienna, Steyr, Prague and Budapest, though the large Škoda armaments works at Pilsen (Plzeň) were especially important. Northern Bohemia was industrialised on a scale almost comparable with German Saxony, with activities ranging from chemicals to textiles and glassware.

Economic Change between the Wars

The spatial pattern of industry established by 1914 had to be adapted to the new political boundaries after the First World War, an adjustment made more difficult by economic depression and a severe attack of economic nationalism in several countries. Despite many changes, however, the pattern in 1939 nevertheless bore a clear imprint of the pre-1914 situation.

Germany had been particularly hard hit. Isolation from raw materials during the war years had encouraged the further development of the already large chemicals industry, with major new plants sited in the Elbe–Saale basin of Central Germany, where there was plentiful process water and electricity generated in lignite-burning power stations. Central Germany also developed through strategic considerations during the 1930s – the big steelworks at Salzgitter and the *Volkswagen* plant at Wolfsburg were the most striking additions. Strategic considerations also led to decentralisation of the Berlin electrical engineering industry to South Germany, whereas after the *Anschluss* with Austria plans for a large steelworks in Franconia were transferred to Linz on the Danube. Division of Upper Silesia between Germany and Poland had made further investment there unattractive; also, the frontier position made investment unattractive in the Saar after its return to Germany in 1935.

Economic depression hit the heavy industrial districts of Northern England and Scotland's Clydeside severely, while growing overseas competition eroded the markets of the cheaper products of the Lancashire cotton industry. In comparison, the Midlands and South-east England blossomed new consumer-oriented industries. Investment in the chemicals industry to offset the near monopoly in several branches held by the Germans pre-1914 had brought new industries to Humberside and even to depressed North-east

England. Decline also arose from technological change – the shift to oil-burning vessels was a major factor in the loss of the high grade steam coal market in South Wales that hit valleys like the Rhondda so hard. Whereas several older iron and steelworks sought to survive as their markets dwindled, attempts to use more home ore, especially the less expensive if low grade scarpland ores of Eastern England, saw the growth of new sites at Corby and Scunthorpe, while to relieve hardship in the South Wales valleys, Ebbw Vale steelworks was opened with government support. In the late 1930s a policy of industrial dispersal under threat of war brought new plants to some depressed areas, such as the big munitions works at Chorley in Lancashire.

With its northern industrial districts and coalfield devastated by war, France tried to extract as much coal as possible from the Saar coalfield until its return to Germany in 1935. At the same time, other sources of energy such as hydroelectricity were developed and encouragement given to industry to develop in less vulnerable locations in the Rhône valley and in western France, but much faith was put in future immunity from invasion through the elaborate Maginot Line defence system, which perhaps discouraged industrial relocation. In Italy, the shift to new engineering products such as motor cars contributed to the foundation of a successful engineering industry, helped and encouraged by the Fascist government that needed it for rearmament. Under Mussolini, a considerable aircraft and shipbuilding industry augmented other branches, while plans were laid to expand iron and steel production, despite the poor natural endowment, though the country's energy needs were to come mostly from electricity generated from several different sources. In Iberia, economic development was arrested in Spain by civil war, while Portugal remained dependent on its traditional agriculturally based industries. Untouched by war, Scandinavia had begun to take advantage of the new technology in metal processing and in engineering to lay the foundations of a high quality, sophisticated industry.

Industrial development under Polish *dirigisme* was mostly well away from the frontier in the south around Kielce, though Polish Upper Silesia's coalmines were linked to the new port of Gdynia by a special railway in the hope of breaking into the Baltic coal market (to the detriment of Britain). In the newly formed Yugoslavia, similar policies sought to revive pre-1914 Austrian plans to develop the long moribund Bosnian iron industry. In Czechoslovakia, Bohemia's industry was badly hit by depression and by protectionist policies closing its markets in other countries, while there was a need to help the Slovaks, whose economy was dislocated by the new frontier with Hungary. One of the main developments was the building of the vast Bata shoe factory at Zlin in Moravia to relieve rural unemployment. Hungary's iron industry had been cut from its sources in Slovakia and the country found itself in tightly drawn boundaries that left traditional markets in other people's territory. Nevertheless, with rich bauxite deposits able to supply Germany's big demand, the Hungarian economy was tied ever closer to that of the Third Reich throughout the 1930s. Austria, stripped of its empire, faced the same problem of an industrial structure too large for its now reduced requirements and desperately dependent on foreign markets.

An immense industrial revolution under way in Russia was laying the foundations in the 1930s for an economic and political link with Eastern Europe 20 years later that was to recast the economic geography of the continent.

The years 1860–1914 had seen emerge, in an economic but also in a social sense, two Europes – one, predominantly in the west, industrial; the other, spread across Eastern and Southern Europe, agricultural. In their contrast they generated a simple but fundamental pattern of trade, though in detail the picture was not so dazzlingly simple. There was, in reality, a number of levels of development and the range depended on factors such as the intensity of innovation, with the industrial and scientific creativeness of Britain and Germany at the top of the hierarchy; the ability to accumulate capital and to invest (in which the Western European countries led); the resource endowment, with the contrast between the success in the coal and iron age of Britain and Germany and the embarrassment of France, poor in coal; the stability of society – the volatility of the Balkans was a major impedance to investors' confidence and to the countries' own ability to concentrate on economic development. No unity of economic philosophy among the countries existed: Britain favoured free trade, though recognising with some reluctance Commonwealth or Imperial Preference; Germany had strong protectionist lobbies and sought to build an economic empire in Danubia and the Balkans; whereas in several of the Succession states after 1918 a powerful *dirigisme* was seen (especially in Poland); and the new Soviet state had sought to exist in a virtually closed system as much by choice as force of circumstance. Natural economic forces, seeking a spatial equilibrium, clashed with governmental desires based on strategic and other needs, so that the political quilt clearly had powerful formative effects on the spatial pattern of economic geography. The strong economic growth – particularly in industry – between 1860 and 1914 had taken place in a political–territorial framework radically changed by the new pattern of states after the First World War, so that the years 1920–1939 were spent trying to adapt the existing pattern to the new demands, while the decline in overseas influence of several of the imperial powers had a further influence on economic geographical patterns in the continent.

The economic geographical pattern of Europe is clearly dynamic and constantly changing in response to new resource appraisal, new technology, new markets and new demands. For example, in Britain the nineteenth century marked the rise of new industrial regions – Scotland, North-east and North-west England, the Midlands – that eclipsed older centres of pre-industrial activity, such as the Forest of Dean, the Weald and the southern woollen towns. Events were similar if less pronounced in Germany as the coalfields outstripped older areas like Berg and Mark. The mediaeval importance of Flanders and Northern Italy had been eclipsed by English enterprise and by the South German towns; nineteenth-century technological advance in Western European iron making outstripped the eighteenth-century prosperity of iron making in Sweden and the Russian Ural, resulting in these regions' ultimate decline. By the First World War, as well as between the wars, several industrial regions in the North of England lost their significance as new industrial trends developed in South-east England and the Midlands, while the

Elbe–Saale basin in Central Germany began to challenge the Ruhr, and Nazi plans were laid for industrialisation in North-west Germany and even in Austria. '*Dirigiste*' Poland sought in the 1930s to develop strategic industries in the Kielce region away from the vulnerable frontiers, just as Yugoslavia planned industrial growth in interior Bosnia. A vast new industrial structure was being laid in the Soviet Union, much however in so-called Asiatic Russia. Impending war only accelerated these changes.

Transport

In inter- and intra-regional relations, transport plays a vital part, linking disparate areas together to realise their full potential. This is particularly meaningful in Europe, where distances are not great and physical obstacles modest, and the patterns of distribution of land and sea have played an important role in transport evolution. Perhaps the high level of interface beween land and sea, in the absence of vast continental interiors (outside the Russian Plains few places are more than 500 km from seawater) but with the deep landward penetration of epicontinental seas, has been a key historical factor in successful economic development through the ease of trade.

Trade and movement along well-defined routes have existed since earliest times, but until recent technological advances rendered it unnecessary, travel was often a series of stages using different media. Land transport was generally of short distance, not uncommonly a portage between waterways. Pack animals passing along crude tracks were used into the Middle Ages and until much later in upland and mountainous country. Carts appeared in Roman times and became important in the late Middle Ages, remaining until motor power displaced horses and oxen. The first true road system was built by the Romans, primarily for strategic purposes though also used for appreciable trade. Lack of effective organisation in post-Roman times hampered its maintenance, let alone its extension. Some mediaeval potentates created their own inferior system of post and military routes. Mediaeval roads were unkempt, crude ways, wide lateral tracts across which traffic shifted to avoid summer dust and winter mud. Although many Roman bridges disappeared, by the tenth century bridge building had begun again. Traffic was small and fluctuating and the importance of routes varied with political fortunes. Mediaeval maps indicate recognised routes like the roads to the Champagne fairs, the Hellweg, Via Emilia, the Brenner – along which were strung route towns, the seats of guilds of professional carriers. Convoys and caravans for mutual safety and help were common, while monastic and clerical communities were a great help to travellers (for example, the Great St Bernard monastery since the thirteenth century). Postal systems set up by some major princes were important ways of communication, though most remarkable was the hereditary postal authority of the princes of Thurn and Taxis for the Holy Roman Empire (established 1490) that lasted in ever-diminishing form until 1866. It is perhaps noteworthy that outside mediaeval Europe were far more impressive systems of communication and travel, as described so wondrously by Marco Polo in thirteenth-century China.

Conditions were generally more backward the further east one went in

Europe. The Crusaders constantly reported the difficulties of crossing the Balkan Peninsula, whereas in Ottoman times little improvement was done, except to build a few strategic bridges, though a mediocre military highway was developed from Istanbul through the Maritsa valley and Morava–Vardar Corridor to Belgrade. Over much of Poland, Lithuania and the Russian princedoms, roads were wide muddy or dusty tracks – as they remain – leading from one fair to the next.

In Western Europe, the seventeenth and eighteenth centuries saw much road building, especially by the French, of long, straight tree-lined roads designed for marching troops – a model copied by Prussia and other states. Reviving economic life from the late seventeenth century brought an increase in carting, often associated with river ports and 'carters' towns'. Prussia built many roads in the eighteenth century and began a major programme after 1816, seeking to link its scattered provinces together. Roads were everywhere brought to areas where they had not existed before, such as the military and subsequent roads in the Scottish Highlands. The particularly vigorous economic development in Britain encouraged the erection of toll roads (turnpikes), which were also built in parts of Central Europe. There was little change in design of road vehicles and methods of traction, so that carts and horse power were overwhelmed by the rising volume of goods to move in some areas, such as late eighteenth-century England, and other solutions – the canal and later the railway – had to be found. The railway was so effective that roads were rapidly eclipsed and little further development of road transport took place, apart from road building in some mountainous parts, through the second half of the century.

The rise of the flexible and versatile motor vehicle in the First World War gave a great impetus to road building between the world wars, especially in Germany and Italy where the motorway modelled on the Roman tradition was developed. In Western Europe a classified hierarchy of motor roads evolved and rising sums of money spent on improvements, though these seldom kept pace with the pressure of traffic growth. In Eastern Europe, the motor vehicle was far less common and motorable roads radiating from towns did not always extend far enough to interconnect. Although the motor car itself was a European invention, the motor car revolution has been one of the few modern innovations brought to Europe from outside, where it had occurred in the early 1920s in the United States of America. Its real effect was to be most felt after the Second World War, again diffusing from Western ·Europe eastwards.

Europe's rivers are mostly of modest dimensions that allowed navigation by simple craft, though they suffer in their natural regime from an increasingly long frozen period eastwards and also from flood and low water (especially in the south) at certain seasons. The less gently graded, faster flowing streams of the mountains and particularly Scandinavia are least useful for navigation, though they were and, to a diminishing extent, still are used for logging. The many remains of simple canoes suggests that rivers were important routes in early times, while there is evidence of a vigorous grain trade between Germania and Britannia along the Rhine even in Roman times. The combination of rivers and short portages was common, as on the historical trade routes from the Baltic to the Mediterranean.

As trade grew so traffic rose, but navigation was hampered by voracious dues levied by princes and by restrictive practices of the towns and their guilds on river traffic, while there was an organisational inability to improve the channel by clearing rocks and shoals. The fortified towns and castles along the Rhine are witness to this period, while great navigational hazards are mirrored in legends like the *Lorelei*. The small, shallow draught vessels were pulled upstream by horses or men, whereas they drifted downstream on the current, which consequently could not be too fast. Funnelling effects in some valleys or steady upstream winds in the plains allowed the use of sailing vessels. On the Rhine and Danube, among others, great rafts were used to transport wood and also other commodities, often manned by several hundred men and needing great navigational skill. River traffic was of course subject to the elements: a slightly faster current in flood, a little fall in draught or the presence of even modest ice brought navigation to a halt. Truly effective action in improving the fairway of rivers began only in the nineteenth century, though some early 'improvements' later had detrimental effects through overstraightening the course – the Rhine in Baden, the Oder and the Tisza were among the rivers to suffer in this fashion. The great revolution in traffic came with the use of steam tugs – first used on the Clyde in 1801, on the Bodensee in 1824, on the Rhine in 1825 and on the Danube in 1827. The steam tug with barge train became the standard means of river traction until the diesel self-propelled barge appeared after 1919.

The Romans built several canals for the carriage of grain and to give easier access to the sea. In the Middle Ages, Charlemagne tried to link the Rhine to the Danube via a canal joining the Main and Altmühl. The first modern canal was a link from the Loire to the Seine in France, opened in 1680, and followed a little later by some canals in the Low Countries. Prussia sought to improve transport by canals that linked together the rivers of the northern plains, often using abandoned meltwater channels. The Bavarians built a canal joining the Rhine and Danube (*Ludwigskanal*) between 1836 and 1842, but it was too small and was quickly eclipsed by railways. The greatest canal-building episode came in Britain, where the rapid rise of industry demanded movement of large bulk freights that carts could not deal with, and the real mania began in 1736 with the Bridgewater Canal and had ended by 1827, when railways began to buy up canals. About 7000 km of canals were built, mainly in the form of a large X-shaped arrangement joining the Mersey, Trent, Severn and Thames basins and intersecting in the Birmingham area, where Europe's only canal town, Stourport, arose. A revival of canal building in Western Europe came in the late nineteenth century, when canals between navigable rivers and to give access to the coalfields were constructed, especially in the Low Countries and the Rhine basin. Britain's canals were unfortunately built only for small, narrow boats but these continental canals mostly carried big barges up to around 1000 tons and could be worked by barge trains. Building continued into the years between the world wars, when the important east–west link, the Mittelland Canal, was opened in Germany, and a link from the Oder to the Danube was begun but not completed, while other projects including the Rhine–Maas Canal and the

Saar–Pfalz Canal were discussed. The Grand Canal d'Alsace, parallel to the Rhine between Strasbourg and Basel, made the latter accessible to barges. The sea-going canals were also significant aids to accessibility – the Manchester Ship Canal (1894), the Kiel Canal (1895), and the North Sea Canal (1865) to give access to Amsterdam – whereas the Suez Canal (1869) in North Africa had important consequences for traffic in the Mediterranean basin.

One of the greatest innovations was the railway that so greatly accelerated movement: the world's first true railway opened in 1825 in North-east England between Stockton and Darlington; in 1830 the more ambitious project from Manchester to Liverpool was completed. During the 1830s many of the first lines in the mainland continental countries were opened, but progress varied from country to country and was generally slower and later towards the east and south. By 1860 the main outline of the railway system in Western Europe was complete and lines were pushing eastwards and southwards – in the 1860s some of the first lines began to appear in the Balkans and there was feverish building in Russia (where the first railway opened in 1838). The period 1870–1900 was a busy time for construction, with the completion of many main and secondary lines. As in Britain, railways on the continental mainland grew piecemeal, short lengths gradually merging into trunk routes. There was no grand design, not surprising in the behavioural environment of the time, and visionaries like Stephenson in Britain and Friedrich List in Germany received little encouragement. By 1900 the density of the railway 'reticule' in any district was a good indicator of its economic health, though certainly there were in some places 'extraneous' factors like strategic needs. Whereas in Britain the railway network had been constructed *after* the main wave of industrial development, the reverse was true on the continent, even in Germany. There was also more state intervention on the continent; consequently fewer duplicating routes were built than by the competing British railway companies, and most towns were satisfied with one station, unlike the three or four of different companies common in British towns. Technical advance also allowed cheaper construction, since it was known by 1840 that much greater gradients could be tolerated than the pioneer engineers in Britain had believed.

Many routes in Europe had great strategic and diplomatic significance: there were powerful political undertones in the St Petersburg–Warsaw–Vienna railway (1861), the vague project of the Berlin–Bagdad railway, or the abortive Hamburg–Venlo–Paris railway, and in many Balkan railway schemes, while the classical diplomatic train became the over-romanticised *Orient Express*. After the 'Railway War' of 1866 between Austria and Prussia, the strategic interest in railways, expectedly less developed in Britain, reached a climax.

The vigorous railway building period ended in 1914 and after the First World War new building in Western Europe was limited mostly to short missing links or lines avoiding junctions. In Scandinavia, it continued with such major projects as the strategic inland railway in Sweden or the Norwegian line northwards to Bodφ and several important routes in Finland. In Eastern Europe building also went on: the Polish coal railway from Upper Silesia to the new port of Gdynia was a good example, but in Yugoslavia, in Bulgaria, in Rumania and even in Albania railways were extended. Even in European Russia there was considerable construction, though not on the massive scale of Siberia and Soviet Central Asia. The latter period of railway building, from about 1880 onwards, had seen some of the most dramatic engineering features to overcome physical obstacles – the great tunnels such as the Gotthard (1882), Simplon (1906), Mt Cenis (1871), Tauern (1909) or the Lötschberg (1911) in the Alps; the great bridges, notably the Forth Bridge (1890), the Moerdijk Bridge in the Netherlands (1871) or the Danish Storstrøm Bridge (1937). Where unbridgeable water obstacles lay in the way, connecting ferries carrying railway vehicles were instituted.

The overwhelming growth of the railways before 1914 left a pattern designed for the political and economic conditions of that time in a new matrix of frontiers after the First World War. This is well seen in Poland, where a dense Prussian railway network typifies the western territories, whereas a broader mesh is found in southern Poland, where Habsburg power had ruled, while in former Russian Poland the reticule remains a sparse mesh. Much of the new building after the First World War comprised short lengths of railway to adjust the system to the new requirements, such as improved links between Bohemia–Moravia and Slovakia, one formerly under Austrian and the other previously under Hungarian rule now united within Czechoslovakia.

With the intricate pattern of coasts and the long fingers of sea reaching deep into the land, sea transport has always played a significant role in trade and inter-regional contact in Europe. In Classical times, Europe's trading relations were played out among the disparate regions of the Mediterranean and, to a lesser degree, the Black Sea. In days of simple navigation, it was possible to sail on these seas for considerable distances seldom out of sight of land, while regularity of winds (for example, the Etesian) and the absence of vigorous currents were a great aid. In the Baltic, winter ice has long been a hazard to shipping, but the fierce waters of the Atlantic and ruthless North Sea, with high winds, sharp seas and howling storms, kept navigation in coastal waters. Early sailing vessels were generally poorly manoeuvrable and of limited seaworthiness, so that even in moderate conditions they had frequently to seek shelter, while the importance of harbours depended on natural shelter and ease of entry and exit under sail. The growth of trade rested not only on demand but also the ability to use the seas, reflected in improvements in vessels and navigation techniques in the fifteenth century that were quickly followed by a widening of the trading horizons. The Portuguese developed the effective sea-going caravel for their long exploratory voyages, whereas in southern waters the carracks and various members of the galleon family were developed (oars were long used in the frequent windstill conditions of the Mediterranean). In northern waters, the Hanseatic cog was perfected as a stable and seaworthy vessel of considerable capacity for its period, though in general Venetian and Ragusan ('argosy') vessels were larger. The Dutch in the sixteenth and early seventeenth century, responding to the growing trade of the Rhine basin, perfected useful commercial as distinct from war ships, but towards the eighteenth century, as Dutch ports silted and demanded shallower draughts that reduced seaworthiness, the initiative passed to the

British and French. A great spurt came with the fast and seaworthy clipper ships developed by the British in the 1830s, though too late as they were quickly overtaken by steam vessels.

Until the coming of railways, almost every small inlet or estuary was a thriving port, while the small size of vessels until the late eighteenth century meant a high level of mobility, but fortunes of harbours depended much on natural conditions. Silting in mediaeval times had claimed several prosperous English channel ports and Chester on the Dee, besides some French harbours, whereas Rattray, a thriving Scottish seaport, completely disappeared when its harbour was closed by a shingle bar. In contrast, Bruges arose as a major port on the Zwin after the flooding of 1134. On the other hand, the Golden Age of Antwerp, aided by changes in the Scheldt channels, declined through political change, while political unrest and the move of the herring from the Baltic were factors in the fall of the great *Hansa*.

Bigger, more complex ships (especially with steam engines) made necessary concentration on fewer harbours, because of depth of water, ease of entry, and facilities for repair and maintenance. Growth of overseas trade, particularly with North America, favoured Atlantic ports such as Liverpool, Cherbourg, Le Havre, but also Hamburg and Bremen, Rotterdam and Amsterdam and Belgian Antwerp. Railways offered fast inter-regional services and small ports declined, except in areas badly served owing to inaccessibility to railways – the Dinaric coast, the Greek peninsulas and islands and the islands and coasts of Norway and Scotland. The specialised port also emerged – the great coal and ore ports (Blyth, Newport, Narvik), the liner ports (Southampton, Bremerhaven, Le Havre), or the ferry ports (Flushing, Dover, Calais, Harwich and the Hoek van Holland). Between the world wars, increasing economic nationalism encouraged the use of one's own ports, reflected in the special railway tariffs offered in Germany to draw trade to Hamburg and Bremen from more natural flows to Dutch Rotterdam, Belgian Antwerp or even Italian Trieste.

Farming

Europe offers a great variety of farming possibilities and there are few large areas where it is completely absent, though in many upland and mountain areas the farmer's life is hard, with an emphasis on pastoralism.[2] Although it is possible to discern regional patterns, the local influence is strong and marked variations take place in quite short distances. Altitude is a significant influence and aspect is also important – with the attraction to sunny south-facing slopes – but angle of slope, even when quite modest, can make arable farming difficult. Soils are another important influence, with parent rock over most of Europe as a notable indicator of their character. Perhaps the most significant influence of all is climate – not simply in terms of precipitation and temperature, but also in terms of insolation through the length of day and the accumulated temperatures totalled from the number of degrees daily above the minimum growing temperature, although the nature of precipitation and type of water supply cannot be ignored.

Northern Europe is distinguished by long winters with short days and summers, short in duration but with long days and surprisingly high temperatures away from the coasts. Precipitation is generally plentiful but not high, though exceeding the low evaporation. Oats, rye, potatoes and flax are main crops, with hay (dried characteristically on poles or temporary fences or racks) for livestock. The cold, acid soils with their raw humus are a serious constraint. Atlantic Europe is marked by its humidity and its lack of extremes in winter and summer. The drier parts are arable country, with potatoes, wheat, oats and barley, but it is more commonly associated with livestock and dairy farming on its lush meadows, whereas the exposed uplands are sheep country. Soils vary much in quality, but tend generally to be moist. Central Europe, with a more extreme winter and warmer summer, is diverse farming country, where relief and aspect are important. Arable farming – grains (wheat on the better soils, rye on the poorer ground), sugar beet, potatoes and now green maize are typical – but livestock is also found, especially in the close relationship between pigs and sugar beet and potatoes. Soils vary from the rich dark, almost black-earths, of the best grain and sugar beet country to the sandy, raw, acidic podzols of the ryelands and uplands. Much of the contrast of productivity and yield arises, however, not from natural conditions but from social and economic contrasts and the quality of farming generally declines east and south-eastwards. Southern Europe is marked notably by tree crops – vines, fruits and particularly the olive. Grain, principally the *pasta* wheats, is also grown, though most widespread in dry uplands (stripped by mediaeval shipbuilders and warring armies of their forest) is pastoralism, with sheep and the destructive goat. In Italy and Spain, large forests referred to in Classical literature are now barren steppe or thorny *macchia*. In the uplands, in pockets of soil and where there is water, rich fruit gardens can be established. Water is the secret of farming in southern Europe, with the long dry summers and wet and sometimes raw winters. The problem is aggravated by large areas of calcareous rocks that readily soak up water, while the soils developed on them are iron hard when dry or cloddy when wet. Irrigation must be undertaken with care for fear of drawing unwanted salts to the surface in the summer heat and high evaporation. All this is made more difficult by unsuitable systems of tenure and by the organisation of society.

Mountain farming has a unity – altitude, aspect and angle of slope are the critical factors whatever the climatic or social conditions. The mountains are traditionally the pastoral areas, whether for sheep, goats or cattle, and most have a history of transhumance – now dead in Scotland, dying in Scandinavia and in the Alps, but still vigorous in Carpathia, in the Balkan and Dinaric mountains, and by no means ended in the Iberian mountains. On the valley floors, hay or arable crops, vines or fruit are raised. In the Dinaric mountains, solution hollows in the calcareous rocks are the only areas with sufficient poor red soils to support cultivation, if not flooded too long by the winter rise in the water table. Soil is often laboriously gathered from crevices to put on the small, stone-walled fields, but the richer basins of the Balkan mountains are renowned for fruit, tobacco and even cotton. In the Alps and Carpathians, cattle are dominant, though there are also plenty of sheep in poorer Carpathia, just as there are in the Iberian mountains, which like the Italian Apennines, the Dinaric and Balkan mountains have both goats and sheep.

Set amid mountains, the Plains of Pannonia east of the Danube enjoy almost steppe-like conditions: in the past they were great cattle country, but later grain became more important, whereas now they have a mixed grain–livestock economy. West of the Danube, in moister conditions, the pattern is more akin to elsewhere in Central Europe. Farming in the plains of Wallachia and Moldavia has a strong grain emphasis, but maize has tended to predominate, being able to feed if not especially nutritionally well the dense peasant population.

Until the nineteenth century, Europe was essentially a continent of peasants. There were nevertheless expectedly regional variations in the ways of farming, the crops raised and the social and tenurial arrangements, but basically farming was on a subsistence basis, for the possibilities of trade in farm produce were limited and the level of productivity low. The separation between the rearing and fattening of cattle meant, however, there was extensive trade in droving. Change and new ideas spread slowly: the three-field system known in Western Europe from at least early mediaeval times was not adopted in parts of Russia until the late fifteenth century, before which time shifting cultivation had been practised. There had been a major mediaeval phase of clearing and spread of cultivation, ended by the devastation of the Black Death (1348), which had seen most of the present villages of Western and Central Europe established, though in places some retreat was later to occur. From the seventeenth century a new phase of rural colonisation began, spreading to parts of Eastern Europe, while great changes in social relations produced further modification in farming from the late eighteenth century, a process still operative.

A two or three open-field system was characteristic over much of Northwest and Central Europe and was found in the Anglo-Saxon settlement area in Britain, while it had also spread eastwards. Enclosure of land for sheep runs in the English Midlands in the Middle Ages has been seen as a factor in the early emergence of distinctive individual farms in compact holdings. In Central Europe, changes in society and tenure were not accompanied by reorganisation into separate holdings and even in our time the landscape remains mostly one of nucleated villages with fields still divided into uneconomic strips. In glaciated parts of North-west Europe, farming on drier ridges, the *Esch*, used an unusual one-field system with elaborate dunging, but population pressure forced the less privileged into the waste and common and contributed to the modern landscape of scattered farmsteads in parts of the Netherlands and North-west Germany. In the lands east of the Elbe and even in Bohemia and Danubia, much land passed into vast entailed estates (usually reducing the peasants to landless labourers), which had from an early date begun to grow grain and later sugar beet or to rear cattle on a commercial scale. The nature of feudalism and serfdom regionally depended commonly on the value of tied labour.

In two areas – Northern Italy and the Netherlands – early development of a vigorous urban and commercial life had considerable impact on farming. In the Netherlands, pressure on land and the rewards of increased production encouraged phases of land reclamation, the success and boldness of the schemes increasing as techniques improved. In Northern Italy, similar incentives encouraged the use of irrigation, while a new element in both areas was investment in farming by the urban *bourgeoisie*, whereas in contrast, Southern Italy remained in the large *latifundia* of Roman origin, with inefficient peasant serf labour. These progressive areas introduced several new crops – in Northern Italy, rice and mulberry; and fodder crops (to aid the keeping of more livestock) and plants for dyestuffs in the Netherlands. Perhaps the most important of all new crops introduced was the potato from the New World, brought by Raleigh to his Irish estates in 1586, whose heavy yield and good food value were later to provide a mainstay in the diet of industrial populations.

Mediterranean farming is carefully adjusted to the long dry summer and the moist winter. Small grains have been the mainstay, but tree crops and the vine (introduced into Central Europe by the Romans) have also been important. The vine had been brought from Persia in prehistoric times. The typical traditional Mediterranean system was a two-field rotation, allowing half the land to be fallow, a not always successful precaution against soil erosion. After the breakdown of Roman law and order, neglect of drainage and increased run-off from the overgrazed uplands resulted in a growing swampiness of some of the best alluvial lowland and its malarial infestation. From earliest times, the olive has been the source of fat, since the lack of summer forage always restricted cattle keeping and butter making. Goat herding was important in reducing the forest area that also led to a decline in pigs masted in the woodland. By mediaeval times, the worst of 'share cropping' and bad landlordism plus soil erosion and population pressure brought a falling productivity that destroyed the rustic image painted by Roman poets.

The Romans had used irrigation on a small scale despite their organisational and engineering skills, but such techniques without adequate central authority proved generally impossible in post-Roman times. The Moors, however, introduced large-scale irrigation into Iberia, where winter snows in the mountains provided spring and summer water, with a shift from traditional grains to more valuable crops; but with the reconquest of Moorish-held lands, many irrigation systems fell into disuse. The Arabs introduced citrus fruits into Europe, and rice first came with the Age of Discovery, as did sugar cane.

In the lowlands of South-eastern Europe – in Moldavia and Wallachia and parts of eastern Hungary – large tracts of land remained underpeopled and even unoccupied until the later eighteenth century, but thereafter became some of the densest settled and most poverty-stricken countryside in Europe. The Turkish invasions drove away Christian peoples to the mountains or forced them into closely nucleated villages for defence. Although disinterested in maximising economic potential, the Turks gathered many peasants into serfdom in large estates (*čifliki*) and only areas near the main towns flourished, notably the Maritsa basin in Bulgaria, well placed to serve the Istanbul markets. Here arose intensive garden farming, but elsewhere farming became extensive as settlement declined in face of Turkish oppression, with a rise in pastoralism as herds could be driven to safety in raids, whereas standing crops were commonly burned.

The mediaeval peasant had depended on the forests for firewood and wood for making many articles, for food and for bedding animals. In the forests, pigs were masted, but from an early date forest had been cleared for colonisation, particularly before the Black Death. Extraction of too much timber for shipbuilding (in Mediterranean forests) or charcoal making (as in the Weald) destroyed extensive areas, while grazing of sheep and goats in the southern European mountains reduced the forest, especially where rejuvenation was slow. It was not until the eighteenth century that an awakening to the value of forests came and there was much replanting, though not always with original or even appropriate species, but the nineteenth century saw development of proper forest management, notably begun in Germany.

Although not without its ups and downs – crop failure, famine, epidemics and other major natural catastrophes – Europe until the nineteenth century had a remarkably balanced economy, dominated extensively by subsistence peasant farming and pastoralism, though from late mediaeval times this had been under increasing pressure from the growing towns and their commercialism. The great discoveries of the fifteenth century were followed by new crops, of which the remarkable American potato was to become the staple diet of Western and Central Europe of both urban and rural populations, just as maize was to support high rural densities in South-eastern Europe. Most important were the new ideas about farming techniques developed in the eighteenth century, notably how to reduce fallow and improve rotations to give better yield, whereas the problem of keeping livestock over winter was eased by development of suitable fodder crops. Selective breeding came to improve the yield of livestock, which was no longer allowed to roam the common, the fallow and the forest. Once again, the new techniques spread from the West, notably from England, where the Improving Movement and the Agricultural Revolution were forerunners of similar trends on the continental mainland. From the mid-seventeenth century, moves to free peasant land-ownership gained ground, though this process again went furthest in Britain. Eastwards, southwards and south-eastwards peasant emancipation was to come ever later, while semi-feudal bondage sometimes lingered despite the original intention of legislation to the contrary. In Eastern Europe, peasant emancipation came only after about 1820 in most areas and was not complete until the 1880s, whereas in Russia the emancipation of 1861 only partly solved the agrarian problem.

Between the world wars, the agrarian problem was to become the key economic dilemma over much of Europe, varying in form and intensity regionally. In the industrialised countries, new economic opportunities allowed people to leave the land and ease rural overpopulation, but at the same time, where trade developed widely, there was serious competition from cheap overseas food, notably true in Britain. Yield and productivity had not, in general, kept pace with new demands and prices became depressed, so impoverishing the farming communities, which had also failed to attract sufficient new investment, especially in buildings and machinery (in Hungary, rural overpopulation became so bad that the government forbade further mechanisation of farming, simply to provide jobs). Some areas, reasonably prosperous in the early to mid-nineteenth century, had become so depressed by interwar times that they formed hearths of malnutrition and deprivation, such as Moldavia and Wallachia. A policy of *laissez-faire* and free trade had been a significant element in the collapse of 'high farming' in Britain about 1870 – by the 1920s farmers selling out were providing cheap land to encourage urban sprawl. In Germany, the powerful Prussian agrarian interests had forced through a policy of high tariffs against imported food and a dear food policy that protected peasants and landowners. Denmark had managed to recast farming to a new market-oriented production of items such as bacon and butter for export, whereas the Dutch exported their butter and ate margarine made from cheap imported vegetable oils. Most Eastern European countries talked of land reform – some even tried it – but the landowners were generally too powerful to allow it to proceed far enough to make an impact, but equally they were reluctant, or more usually unable, to invest in industry or other measures to ease social and economic tensions. In Southern Europe, the old ills were perpetuated – lack of investment in farming by absentee landlords, continuation of unfortunate tenurial arrangements, and an inability to diversify opportunity. The small Baltic countries succeeded in exporting foodstuffs to Western Europe. In the Soviet Union, a massive and ill-conceived experiment in collectivisation was to be spread, after 1945, widely into Eastern and South-eastern Europe. The approach of war in the late 1930s and the relation of food supply to total war brought a sober reassessment of the sorry state of farming in most potential belligerents.

References

1. Useful studies of the development of trade and industry are C. T. Smith, *An Historical Geography of Europe before 1800*, Longmans, London (1967) and W. G. East, *An Historical Geography of Europe*, Methuen, London (1966).
2. A useful source is B. H. Slicher van Bath, *Agrarian History of Europe* A.D. 500–1850, Edward Arnold, London (1963).

6 The Continent after 1945: A New Europe Emerges

Europe emerged from the Second World War more shattered than from the 1914–1918 war, to the aftermath of which it had hardly properly adjusted before the new conflict began. Unlike 1919 no grand peace conference had been held to confirm the new alignments and it was not until the early 1950s that the true impact of the war became apparent. Although the political map showed less upheaval than in 1918, when so many new states had appeared, vital changes had taken place that signalled the ultimate end of the European Age. There was a steady disintegration of Europe's overseas empires, whereas marked influence was established in Europe by peripheral great powers – the Soviet Union and the United States. Much of Eastern and Central Europe was swept into a new Soviet imperium by the Red Army, whereas the United States came to exercise great influence among Western European states that sought to hold off any further Soviet advance. The rapid emergence of a Western and Eastern *bloc* with a deep social, political and economic divide came to a confrontation of stalemate along Winston Churchill's Iron Curtain, cutting through the heart of the century-old concept of *Mitteleuropa* that now evaporated into the vacuum left by the collapse of the German *Reich*.

Territorial Problems

Perhaps the least expected outcome of the Second World War was the creation of an immense Soviet sphere of influence through military occupation that brought a social, political and economic revolution, encapsulating much of Eastern, South-eastern and East Central Europe. Governments favourable to and in some cases dependent on the Soviet Union were set up in Poland, Hungary, Bulgaria, Albania and Yugoslavia, quickly followed by takeovers in Czechoslovakia and Rumania, while the Soviet Occupation Zone in Germany modelled on Soviet principles was eventually turned into the German Democratic Republic. Misunderstanding as much as bad faith or deliberate policy brought a quick collapse of the superficial wartime unity between the Western Allies and the Soviet Union, leading to dangerous confrontation and breakdown of easy contact in the years 1946–1950 – the so-called *Cold War*. By the time the rift began to heal, two new spatial patterns had been established.

Following Allied agreements before the great split, all German territory that lay, at the end of 1937, east of the river Oder and the western Neisse passed to Poland and the Soviet Union, though a definitive frontier was to be fixed only in a final German peace treaty. Poland was compensated with rich and well-developed German territory for its own losses of poor and under-developed country to the Soviet Union east of the approximate line suggested by Lord Curzon, in the First World War, as a possible Polish–Russian frontier. Russia gained the strategically important part of German East Prussia around Königsberg (Kaliningrad), besides retaining the small Baltic coun-tries – Lithuania, Latvia and Estonia – and strategic Finnish territory around Lake Ladoga and in the far north, where there were useful mineral resources, so that a common Soviet frontier with Norway now arose. In the south, Volhynia and the Polish town of Lvov became Russian as well as the eastern tip of Czechoslovakia, with its useful Carpathian passes and the outwash fans on the edge of the Mid-Danubian basin. Rumania lost Bukovina and Bessa-rabia, bringing a Soviet hold to the northern flank of the Danube delta. The Soviet Union had thus come to have an important strategic hold over the Eastern European roots of the main continental peninsula, while the Soviet occupation zone in Germany brought the Red Army to within 50 km of the lower Elbe and a vital westward salient in Thüringen pushed it towards the Rhine.

Territorial change was otherwise small, though Germany was stripped of its other territorial gains made after 1937, with Austria once again created an independent if occupied state. The Netherlands, Belgium and Luxemburg had small modifications to their boundary with Germany, but the German–Danish border question in South Schleswig was not reopened despite pressures. A significant role in postwar developments was, however, played by the inclusion of an extended Saar territory in the economic fabric of France until 1957, when it returned to Federal Germany. Italy lost small terri-tories on the Adriatic coast, particularly the Istrian Peninsula, to Yugoslavia and Italian islands in the Aegean went to Greece, whereas wartime gains in Yugoslavia and Albania, occupied in 1939, were returned. Bulgaria lost terri-tory in Dobruja to Rumania, which also regained from Hungary its Transyl-vanian lands. A Yugoslav claim on ethnic grounds to Austria's Klagenfurt basin was, however, rejected.

Whereas the four wartime Allies were able to agree to the reinstatement of an Austrian nation state and to remove their occupation forces in 1955, the situation in Germany produced quite different results. The division between the Allies increasingly separated the three western occupation zones from the Soviet Zone as political and economic policies diverged. The Western Allies went ahead to form a West German state, a federal republic of eleven *Länder*, and the Soviet authorities turned their zone into a republic modelled on the socialist *bloc* peoples' democracies. The four-power status of Berlin was chal-lenged unsuccessfully by the Soviet Union, when the 1949 blockade of the western sectors of the city was defeated by a year-long airlift. Berlin, neverthe-less remains a divided city: the western sectors remain isolated not only from the eastern part of the city, but also from the countryside around. Widely dis-parate in population and natural resources, let alone in their social and econ-omic systems and political structures, the two new German 'national' states may be seen as the major innovation of the map of post-1945 Europe. Their long-term future remains one of Europe's most intractable problems.

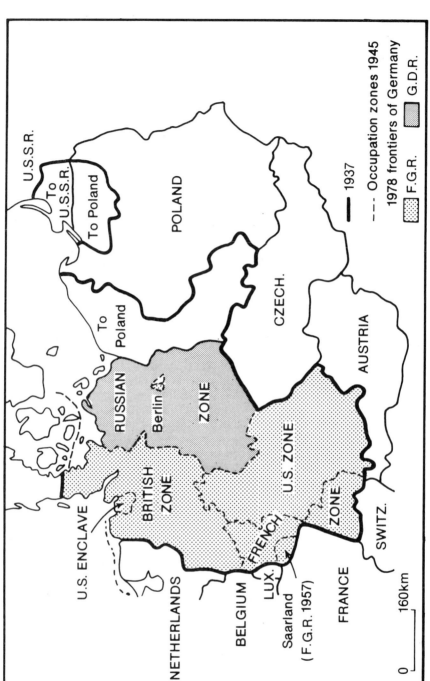

Figure 6.1   Changing frontiers of Germany, 1939–1978

A New Europe

The Cold War generated by the ideological split in Europe had thus caused a division of unparalleled magnitude, separating two political *blocs* that encouraged new social and economic alignments in a recasting of traditional regional patterns. The differences in approach of the centrally planned socialist *bloc* economies and the free market economies of the Western *bloc* have in particular exerted considerable influence on the new spatial patterns. It also raised a fundamentally new approach to European unity. These radical changes were complicated by the generation of a technological revolution as the new spatial alignments germinated, which was to change further the requirements of industrial patterns and transport systems.

The first postwar task had been to get Europe back on the road towards peacetime conditions, though this was made increasingly difficult as the tension of the Cold War mounted. The Soviet problem was relatively straightforward: its new found military and economic power had to be backed by an adequately strong economic base, particularly if the pressures, real and imagined, exerted by the Western *bloc* were to be withstood, especially those generated by the technology gap. For this purpose the new satellite states were harnessed to the Soviet economy in every possible way, often to the detriment of their own interests. In the Western *bloc*, the main fear was of a further spread of Soviet political influence, which it was believed fed on privation and hunger. It was therefore generally considered vital to raise living standards, though this could only be done with care, because any excessive release of pent-up consumer demand would have led to serious economic overheating and self-defeating consequences. To overcome the problems posed by rehabilitation, co-operation and united action were essential, since the countries of the Western *bloc* individually had inadequate resources to tackle the challenges alone. Clearly the economic nationalism of prewar times was to be rejected.

With their economic potential little affected by war, the Americans offered considerable aid for European rehabilitation, following the proven value of United Nations' Relief and Rehabilitation Agency assistance. This assistance, both material and financial, was made through the Marshall Aid programme in 1947, administered by the Organisation for European Economic Co-operation, a United Nations agency in Paris, and a first step in the international approach, later to become the more widely based Organisation for Economic Co-operation and Development. Whereas the Western European states joined the Marshall Aid programme wholeheartedly, attempts by Poland and Czechoslovakia were thwarted by the Soviet Union, which had itself felt unable to join on the basis of its own concepts of sovereignty, though in reality no doubt other fundamental ideological issues were also pressing.

Some early attempts to get international co-operation were expressed in movements towards customs unions, almost unthinkable before 1945. A proposed union between France and Italy proved abortive, but a successful union between the Netherlands, Belgium and Luxemburg (Benelux) was created in 1948 and pointed a way ahead. A somewhat idealistic concept was followed by Western European countries in the Council of Europe (1949), but real international co-operation came in a particularly critical sector. In the early postwar years, Western Europe had a serious problem of energy supply, still derived chiefly from coal, and also a pressing need for more steel for reconstruction of industrial and transport installations. The production of both coal and iron and steel had an uneven distribution among the Western European countries and, to rationalise production and free movement of raw materials and finished goods alike, some overall co-operation appeared necessary. It was also important to 'europeanise' these sectors, because the largest supplier, the coalfields and iron-making districts of West Germany, could best be drawn back into the community of nations through suitable guarantees of supervision and without fear of long-term German designs. With the German Federal Republic established in 1949, the International Ruhr Authority created to regulate German heavy industry became less appropriate to control coal and steel production. In 1951, the first organisation with some, if limited, measure of supranational authority, the European Coal and Steel Community, came into being. Unfortunately it did not attract as wide a membership as was hoped and was joined only by France, Italy, West Germany and the three Benelux countries. Although it helped substantially to ease the problem of demand having outgrown production, it quickly demonstrated that an approach to European integration on a sector-by-sector basis was not appropriate. Its success was possibly more through the circumstances at its foundation rather than through its own concepts, for within a short time serious structural crises in the coal and steel industries began to develop. Nevertheless, its initial success had encouraged members to consider a wider supranational organisation aimed at far-reaching political and monetary union, which American policy was pledged to aid but was unclear as to how it might do so. The outcome was that the six countries of the ECSC came together in the European Economic Community, whose aims were expressed in the formal conditions of the Treaty of Rome (1957) as broadening a customs union into major financial and political integration accepting supranational authority. Britain, Scandinavia and some small Western European states preferred to remain in a looser association pledged to free trade until they felt able to join the wider aims of the European Economic Community, and consequently founded the European Free Trade Association (EFTA).

In his declining years, Stalin's xenophobia had increasingly forced the Soviet Union and its satellites in a closed economic system. The attempt of Poland and Czechoslovakia to join the Marshall Aid programme had been countered by the formation of *Comecon* in 1949, sponsored by the Soviet Union and claiming as its first members Poland, Czechoslovakia, Hungary, Rumania and Bulgaria, shortly joined by the German Democratic Republic and Albania. Yugoslavia, already split from the mainstream of the Socialist *bloc*, remained a fence-sitter. The first six years of *Comecon* were shrouded in mystery, but by the mid-1950s economic and social achievements in Western Europe were forcing it to advertise itself and offer some counter-balancing achievements.

The spatial pattern of economic development was therefore to be much

Figure 6.2 Membership of selected international organisations: (A) EEC; (B) EFTA; (C) *Comecon;* (D) associate member EEC; (E) associate member EFTA; (F) NATO; (G) Warsaw Pact; (H) unaligned, neutral, etc.

influenced by the division of Europe arising from the creation of this new Soviet imperium as a largely closed economic system, with the disintegration of Germany and the evaporation of *Mitteleuropa*. Equally influential was the emergence of the new internationalism in Western Europe, as was the decline of European empires overseas; but one of the most important factors was the passing of economic impetus to North America and a subsequent penetration of Europe by North American economic interests.

New Trends in Economic Geography

The importance of coal as a major energy source and the need for iron and steel for reconstruction at first influenced rehabilitation in both *blocs*. In Western Europe, it greatly affected Allied policy in Germany by the need to get the Ruhr coalfield into the fullest production, and the rising need for iron and steel undoubtedly led to an easing of the levels of industrial production allowed in the western occupation zones. It was likewise the same problem that pressed France to claim the economic incorporation of the Saar coalfield and its steelworks. Coal also provided some of the initial contact with the new Soviet closed system economy, as Poland was able to replace Britain in several markets with coal from its newly acquired mines in Upper Silesia and Waldenburg (Walbrzych), which began to flow in considerable amounts to Scandinavia and to Italy. The need for coal also encouraged Belgian development of the northern Kempenland field as the mines in the older Meuse trough declined, and the thought of enlarged coal reserves spurred Dutch claims for territory along Germany's border. Undergoing a radical reorganisation and needing a fundamental modernisation, the British coalmines were unable to supply exports on a major scale, partly through their weakened productivity and partly through their high cost. There was an intense search for coal: possible new seams were discovered in Britain, bituminous coal was located at great depth in the North European Plain in the German Democratic Republic; additional resources were mapped in Poland and in western Russia; while output and exploration were increased in Southeastern Europe. Output was constantly expanded, but by the mid-1950s the growing use of oil in Western Europe produced a sudden overproduction of coal and consequently structural crises in several fields, though perhaps most dramatically in the Ruhr coalfield. Two factors, however, prevented an even more drastic fall in demand for coal: the rising demand for electricity, where coal-fired power stations were important, and a continued high demand for metallurgical coke from the iron and steel industry, even though the ratio of coal in the raw materials input declined. In Western Europe, many small fields have been abandoned, even though workable reserves in them remain. There has been, as yet, no corresponding trend in Eastern Europe, where the only criterion for abandonment has been exhaustion.

Before 1939 Europe, particularly the more industrial nations of the West, had begun to import increasing quantities of petroleum, much coming as refined products from overseas refineries near the fields. Strategic fears drove the Germans in the 1930s to develop a considerable synthetic petroleum industry and like many other states they began an intense search for reserves in

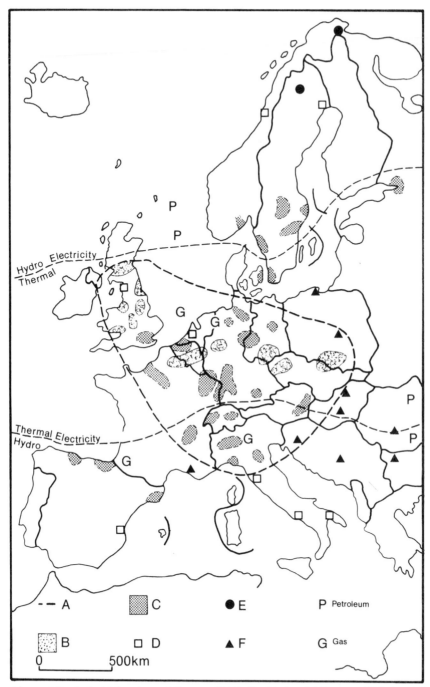

Figure 6.3 Industrial regions of Europe: (A) limit of industrial Europe; (B) industrial region based on coal; (C) industrial regions; (D) coastal iron and steel plant outside an industrial region; (E) major iron-ore source outside an industrial region; (F) emerging industrial region

their own territory, with mediocre success. Wartime demands rose sharply, despite rigorous civilian rationing, and after 1945 civilian demand soon began to rise again. The explosion of motor car ownership in the 1950s in Western Europe was encouraged by falling petroleum prices that also accelerated the use of oil as an industrial fuel, particularly in thermal electric power stations. There was also a rapid development of petrochemicals to form the basis of a widening range of synthetic substances. Threats from overseas countries to nationalise their oil industries, plus the problems of paying for oil in hard currency, as well as the growing sophistication of refining techniques, encouraged relocation of refining capacity from overseas fields to Western Europe. To ship crude oil was cheaper and easier than refined products; the refining could be done near to the consumer and could take advantage of the latest developments in by-products, whereas there was a freedom to choose the cheapest sources of crude oil and to save hard currency. The process was eased as naval architecture was able to design ever larger tankers with correspondingly lower unit costs, though the bigger the ships became, the fewer the ports at which they might discharge.

In Western Europe, oil refineries show markedly coastal location, dominated by the massive agglomeration at Rotterdam-Europoort. By the mid-1960s several sites were no longer optimal, because they could not be reached by the supertankers giving the cheapest transport costs. The crisis of 1956 that closed the Suez Canal had given a stimulus to growth of large tankers able to carry Persian Gulf crude oil cheaply to European markets around the Cape of Good Hope. In the 1950s distribution inland was mostly by rail and barge, but the mounting demand encouraged pipeline development, especially to the growing German inland markets. There subsequently developed the piping of crude petroleum from coastal terminals to inland refineries, particularly focused in the Rhine basin.

Coal has held a larger share of energy generation in the Soviet *bloc* than in Western Europe. Some consumers and certain special metallurgical plants in Eastern Europe are largely dependent on Soviet coal. Only in the German Democratic Republic has bituminous coal output decreased, the result of exhaustion of the seams. An important factor has been the absence of the motor car explosion in the socialist *bloc* countries, but certainly Soviet experience of a rich endowment with coal reserves and easier technological problems of using coal compared with petroleum has influenced policy in Eastern Europe. The structural crisis experienced in the West European coalfields in the 1950s had no counterpart in Eastern Europe. Like Western Europe, there has been a general but not very rewarding search for petroleum – in Rumania, once a major world producer, new deposits have offset exhaustion of older fields but without a massive increase in production. Only the Soviet Union has registered large new deposits, notably in Western Siberia. A rising demand for petroleum in Eastern Europe has been largely met by the building of pipelines from the Soviet Ural–Volga fields to Poland, the G.D.R., Czechoslovakia and Hungary. Some crude oil is imported into the G.D.R. via Rostock from Third World trading partners, and the Danube basin will be supplied with crude oil from similar origins via a pipeline from the Yugoslav oil terminal at Rijeka-Bakar. Crude petroleum is, however, a valuable export commodity to the capitalist world for the Soviet Union, which has told its Eastern Europe neighbours that they cannot expect to remain first priority. A Soviet oil terminal is being built at Ventspils on the Baltic.

The growth in the use of petroleum also encouraged greater interest in natural gas, which was developed on a considerable scale in the Soviet Union during the 1950s. Especially important for Western Europe was the discovery of considerable resources in the North European Plain, particularly in Groningen province in the Netherlands. Dutch natural gas, now also supplied to neighbouring countries, has greatly reduced home demand for coal, so much so that little interest is now shown in the Netherlands' own native coal resources. The discoveries in the northern plains encouraged a search in the southern North Sea, where the most fruitful finds in the British sector have much eased the British gas supply position at the expense of the coal industry. Finds have been made in Eastern Europe, but not on a vast scale, though Rumania has sufficient gas to supply itself and northern Hungary. The Soviet Union is likely to become a major natural gas supplier to Western Europe; already it is sending gas to Western Germany and to Italy, extending a gas pipeline system that delivers to Eastern European *Comecon* members. Experiments have also shown it possible to liquefy natural gas for tanker transport, and shipments are now made from North African fields.

The impact of technology is seen nowhere better than in the search for oil and gas, particularly in the extremely difficult conditions of the northern North Sea. Great advances in seismic and gravimetric survey have been coupled to computer techniques and to a completely new technology of undersea drilling as well as entirely new diving techniques. The immense and expensive effort has been possible only on a multinational scale, aided by rising world oil prices that have helped to offset the high costs. The major oil discoveries to date (1978) have been primarily in the British and Norwegian sectors, but exploration in the Celtic Sea (British–Irish sector) is also promising. Finds already made are sufficient to change radically the British energy position (and equally its financial position) and to have a heavy impact on the energy situation in the European Community, though widely varying estimates have been made of how long the reserves will remain worth exploiting.

Electricity has become an increasingly important source of energy and the biggest expansion has been in thermal power stations using coal, lignite, oil, natural gas or even nuclear power. The first applications of nuclear power had been in the United States and in Britain, but inauguration of *Euratom* in 1957 began its development in Western Europe. In the European Community, only France and Germany mine radioactive materials, but the largest generating capacity remains in the United Kingdom. Although Germany was slow to join the development of power stations, a large construction programme is now under way. Initially the Soviet *bloc* had to catch up technologically, but by the mid-1950s the first small generating plant was working in the Soviet Union and help has been given to Eastern European *Comecon* members to develop their own stations. The availability of nuclear fuels in the Soviet *bloc* is good, but with generous endowment of conventional fuels, nuclear power is probably less important for the Soviet Union than for its

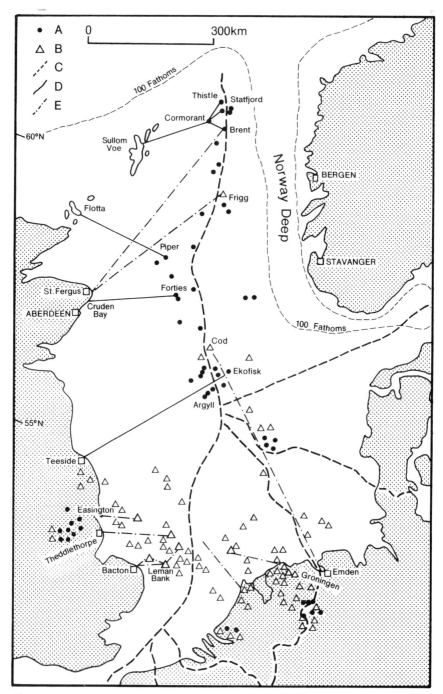

Figure 6.4   North Sea oil and gas exploitation. A oil fields; B gas fields; C median line and international frontiers; D submarine oil pipeline; E submarine gas pipeline

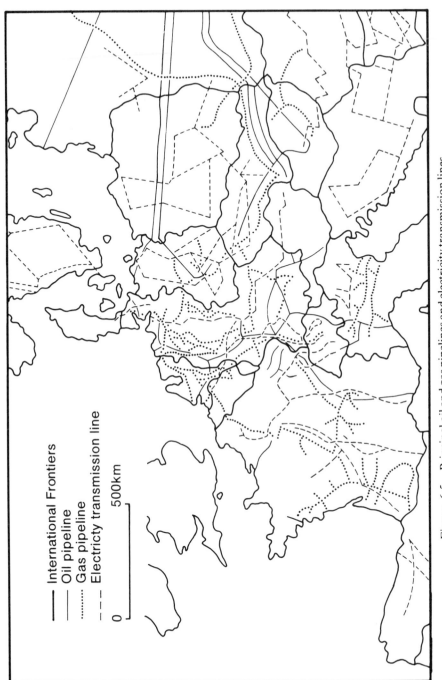

International Frontiers
Oil pipeline
Gas pipeline
Electricty transmission line

0   500km

Figure 6.5   Principal oil and gas pipelines and electricity transmission lines

Eastern European neighbours. Whereas in Scandinavia and in the Alpine countries, hydroelectric power was important before 1945, its development has remained limited elsewhere and in some countries (for example, in West Germany and the United Kingdom) its share in generating capacity has actually declined. Growth of electric power generation has been given an especially important priority in Eastern Europe, where large numbers of new thermal stations have been built, but there are also some significant hydroelectric generators (based largely on Soviet technical experience). Rumania and Bulgaria have given particular attention to hydroelectricity, and a large project on the Danube at the Iron Gates has been jointly financed by Rumania and Yugoslavia. Within *Comecon*, the most important development has been the linking together of the member countries by an efficient grid system which did not effectively exist before 1945 outside then German territory and Bohemia–Moravia.

As demand is likely to continue to rise, energy supply will remain a major European problem in both the Eastern and Western *blocs*. The problem has been confounded by the energy crisis of 1973 which highlighted Western European dependence on extra-European sources. Rising petroleum prices and the insecurity of supplies from volatile political areas of the world intensified interest in native sources, such as the oil and gas of the North Sea, and in alternative sources, such as nuclear power or a new role for coal. Following the model set by petroleum exploration, it has been suggested that remote controlled working of sub-sea coal resources of the North Sea might be investigated. Higher world prices have encouraged sales of Soviet petroleum and natural gas in Western markets, so reducing the priority enjoyed to date by Eastern European *Comecon* members in Soviet supplies, and rising world prices have resulted in some 'imported inflation' into these otherwise insulated economies.

The influence of technology on economic geographical patterns is well illustrated by the iron and steel industry, where new techniques in the industry itself and in related transport have been changing the established locational needs. Immediately postwar world shortages of iron and steel production had brought massive investment in the 1950s, including the building of steelworks as key plants in the concept of autarky in the Eastern *bloc* countries. With new techniques bringing greatly enhanced productivity and a fall-off in the growth of demand, by the 1970s there was an apparent serious overcapacity, particularly in older and less competitive plants, despite rationalisation programmes. Three worldwide factors have influenced the European iron and steel industries: the increase in the number of national industries (1938, less than 20; 1975, over 80 – with Albania and Bulgaria as new members in Europe) as newly independent countries have sought to industrialise; the improvement in technology enabling the ratio of raw material input to finished product output to be reduced, accompanied by a speeding up of the processes; and finally, the revolution introduced in ocean transport by the large bulk carrier, with a consequent sharp fall in unit freight costs. This latter development has been significant in increasing the range of raw materials sources, but has put a premium on waterside sites. The combination of new technology and the bulk carrier has raised the threshold of economy of scale into ever larger works – the optimal annual capacity has risen from about 1 million tons in 1947 to about 10 million tons in 1974. For Europe, these changes have meant a fall in its relative world position, with the United States, Russia and Japan coming to occupy the principal roles: of significance has been Germany's slip down the table.

The classic steelworks has become the Japanese waterside works of around 8–10 million tons capacity fed by bulk carriers. Among the European countries Britain has best been able to take advantage of this location, but structural problems in the industry and all-too-typical labour problems have resulted in a disappointing performance. Large new plants of this type have been established in South Wales, the North-east of England and proposed for the West of Scotland, though within a European Community situation it might be reasonable to expect a much greater expansion. In the past, government policy of forcing the use of high cost home-produced inland coking coals has also been a deterrent to success. Both France and Italy have built waterside plants: Italy is particularly dependent on imported raw materials, but its southern works cashes in on the large cheap labour supply of the *Mezzogiorno*. German plans to build a waterside works at the Elbe estuary have not yet (1978) materialised and the existing Bremen plant is too far upstream to be developed effectively. The Ruhr industry has depended on improved water-borne transport of ore from Rotterdam-Europoort, with works increasingly concentrated on the Rhine frontage, whereas the eastern Ruhr has developed processing of crude steel from the Dutch waterside works at Ijmuiden. The works of the new regional concept of *Saarlorlux*, it is claimed, are sufficiently inland to remain competitive in interior markets. Everywhere, however, inland producers in Europe have tended to become more specialised in the finishing branches and to depend more on pig iron and crude steel from coastal works. Inland producers have also been important innovators; for example, the Austrian industry, greatly expanded by the Germans in the late 1930s, developed the significant oxygen steel process (LD-steel), whereas the Swedish industry has survived on quality steels with the special characteristics imparted by electrical processes.

Stalinist economists had pressed autarkic policies on the Soviet Union in the 1930s, in which iron and steel had received a high priority. When these concepts were translated into Eastern Europe in 1946–1950, iron and steelworks were also accorded priority. In the early 1950s a rash of new works appeared, often apparently without thought to the question of domestic market size and raw materials supply, though it was to Soviet advantage for political–geographical reasons to supply coal and ore. This was typical of the German Democratic Republic, which prewar had depended on works now in West Germany and Poland and was peculiarly ill-supplied with raw materials. In Poland, spread of industry as part of regional policy brought the large new Nowa Huta works east of Kraków; in Hungary, the Dunaujváros plant on the Danube was tied to Soviet and Bulgarian ore and local coal; to even out regional development in Czechoslovakia, a large plant entirely dependent on Soviet raw materials was built in eastern Slovakia. Similar patterns were followed in Rumania and in Yugoslavia, whereas Bulgaria and Albania each planned an entirely new iron and steel industry. When the idea of autarky was

wearing thin in the 1960s and more effective policies were being formulated in
Comecon, these works had become an embarrassment.

New technology brought a large new product range and vigorous growth in
the chemicals industry, particularly through the development of petrochemicals. Marked areas of growth have been the Rhine axis and the Rhine–Maas
delta, where petroleum refining was growing, but the French coasts and
southern England have attracted similar developments, while non-petroleum
based processes expanded on Humberside and North-east England. The early
1960s marked expansion of chemicals in Eastern Europe following a Soviet
decision that their own industry was backward and underdeveloped and the
growing output of petroleum, especially in the Volga basin, made development attractive. In Eastern Europe, the main focus of growth has been in
expanding and modernising the vast chemicals plants of the Elbe—Saale
basin in the G.D.R., but there has been growth in Bohemia and Poland has
used prewar German plants taken in Silesia to expand. Bulgaria, Rumania
and Hungary have also received aid from other *Comecon* countries to develop, with particular emphasis on fertilisers and agricultural chemicals.

The postwar period—particularly from the 1960s onwards—brought a
market revolution in Western Europe as the demand for sophisticated consumer durables rose. In Western Europe, the private consumer has come to
dominate the market, but in Eastern Europe planned investment priorities
have given the consumer a low rating. Consumer durables are a footloose
congeries of assembly industries, mostly closely interlinked, in which labour
is an appreciable element in cost and an important attractive factor in location. With large reserves of labour available from the *Mezzogiorno*, Italy became in the 1950s one of the major producers of domestic electrical goods.
Taking up the initial underemployed expellee population, West Germany had
also developed quickly in the 1950s only to find it necessary to use foreign
immigrant labour a decade later. From the late 1960s, the ever-rising labour
costs in Western Europe began to force producers to examine possibilities of
using low cost labour in Iberia and southern Europe and even beyond (for
example, assembly of cameras and electrical goods by European firms in
Singapore, Taiwan, etc., a pattern recently forced on the Japanese through
similar trends). Consumer durables were slower to develop in Eastern
Europe, despite a fund of labour displaced by reorganisation in agriculture in
several countries. Nevertheless, some link-ups have been established; for
example, G.D.R. electronic parts are assembled by plentiful labour in Hungary. Poland has even considered assembling parts for Western European
manufacturers as a means of earning hard currency.

A characteristic change has been seen in motor vehicle manufacture, the
growth industry of the 1960s and a classic assembly industry in Western
Europe. The first postwar plants were simply expanded prewar locations, but
from the 1960s new plants were located notably in areas needing economic
stimulus, (for example, Merseyside and Scotland in the United Kingdom;
Brittany in France; the Ruhr and Saar coalfields in Germany). The industry
producing quality vehicles also grew in Sweden and even for small cars in the
Netherlands, whereas the multinational nature of several manufacturers
(particularly the penetration by American firms) brought tie-ups between

plants in different countries. Assembly work has also been established in countries with plenty of available labour, such as Spain, and further expansion is likely to be in such areas with copious and relatively cheap labour. The industry has been on a much smaller scale in Eastern Europe: in the G.D.R. dismantling for reparations left capacity less than prewar and no recent expansion has taken place. In the early autarkic period several countries (Poland and Hungary, for example) tried to cover their own needs, but the domestic market was too small for the necessary economies of scale; this was followed by *Comecon* agreements for countries to specialise in a given range of vehicles and to trade by exchange to cover their needs. Poland and Rumania have started building foreign vehicles under licence (also done in Yugoslavia), but the considerable indigenous prewar industry in Czechoslovakia has been slow to expand. Eastern Europe still depends substantially on the Soviet Union for its motor vehicle (both native and licensed designs) requirements.

The new international approach to economic organisation has been reflected in Western Europe by new business trends. Firms have commonly established branch plants outside their own parent national territory, often designed to gain access to untapped labour reserves, or to penetrate wider markets, or even to overcome legal and fiscal barriers. Greater co-operation between firms on an international basis has been seen, particularly for example in the *Saarlorlux* link-up of heavy industry, with joint Franco-German ventures in chemicals and heavy industry: such joint ventures have also sought to rationalise production or to take advantage of new locational trends, like the merger between Dortmund's steel industry and the Dutch ironworks at Ijmuiden. A most venturesome project has been the joint British-French *Concorde* programme, closely rivalled by the European Airbus scheme. Joint European action in the sphere of aerospace development appears the only way in which the excessively large American share of this market can be challenged. The pace for such development has been set to a large measure by the penetration of American capital – for example, Ford of Europe, with plants in Britain, Belgium and Germany or the ill-fated Chrysler–Rootes–Simca link up. American worldwide research and development facilities have been a major factor in the domination of the electronics industry in Europe. In his penetrating study *Le Défi Americain*, Servan-Schreiber[1] pointed to the fact that United States-financed industry in Europe is possibly the third major world producer. The large multinational firms have been generally footloose, so long as their units fit into a closely integrated production system, and they have consequently played an important role in development areas where numerous fiscal advantages could be gained. Consequently the European pattern of some key industries – for example, petroleum refining and distribution – has to be seen in the context of the worldwide activities of their owners.

International companies in Eastern Europe began with joint Soviet companies, notably in Hungary and Rumania, and developed into such organisations as the joint Polish–Hungarian coal-marketing organisation, *Haldex*, whereas later the G.D.R. and Hungary also agreed to assemble German-made electronic parts in labour-rich Puszta towns. Since the early 1970s larger units and wider international co-operation similar to Western multi-national companies have begun to appear through the so-called Socialist economic integration, which should lead to joint industrial products to effect economies of scale. A first project has been a G.D.R.–Polish–Czechoslovak study to pool research, development and manufacturing costs for motor vehicles.

The Transport Revolution

Europe has been swept along by the world revolution in transport. Over most of the continent railways have remained the middle-distance carrier and have made strong efforts to compete with other hauliers through modernisation (electrification, computerisation, etc.), though their place has been more assured in the centrally planned Eastern European socialist economies than in the Western European free market economies. In most Western European countries, railway route length has declined as lightly used links have been abandoned: the trend started in France and Britain and by the early 1970s had begun to affect West Germany. Little new route has been built in Western Europe, though the railway was completed to Bodø in northern Norway and some missing links have been completed in Finland; elsewhere only short new tracks have been added to ease bottlenecks. In their important central position in the European railway system, the West Germans plan some new super trunk routes, modelled on the Japanese Tokaido system, to ease through links. A major blow to the expansion of Western European railway traffic was the short-sighted British decision to cancel the Channel Tunnel. Though long-distance freight traffic has grown, short-distance hauls have gone increasingly to road haulage: passenger traffic growth has been in medium-distance and commuter hauls. Air competition has killed many famous long-distance international trains such as the Orient Express.

Eastern European railways remained almost unchallenged well into the 1960s and in South-eastern Europe considerable new route has been added – including the major Belgrade–Bar trunk line in Yugoslavia, several railway routes in Albania, and numerous lines in Bulgaria and Rumania. New political–geographical alignments have also led to building (for example, the proposed Berlin–Rostock trunk in the G.D.R. and several missing inter-regional links in Poland). Under Soviet influence the railways have been designed mainly for heavy freights and systems of change-of-gauge wagons developed to allow through running between standard gauge route in Eastern Europe and Soviet broad gauge track. As a consequence an elaborate sleeping car network radiating from Moscow has spread into Eastern and even Western Europe.

Western Europe, since 1945, has seen a major explosion in road traffic, with a rather less explosive but nevertheless vigorous development starting in Eastern Europe in the late 1950s. Immense investment in Western Europe has created an extensive system of motorways, particularly dense in the Rhine basin and in Italy and least developed in Iberia and Scandinavia. The road-building programme has also brought the construction of a number of road tunnels through the Alps. New roads have been a significant factor in the spread of footloose industries into rural areas. The sudden mushrooming in the number of motor vehicles has caused acute congestion in urban areas, where investment in traffic engineering has been exceptionally high.

Few new motorways have appeared in Eastern Europe, though a number of modest projects were begun in the 1960s. In South-eastern Europe, construction of simple roads in rural areas accessible by motor vehicles has been necessary. Nevertheless road haulage has been growing and in some countries (such as the G.D.R., Czechoslovakia and parts of Poland), extensive bus services have taken over from lightly used railways. There has been a major growth in road haulage of perishable foodstuffs from South-eastern Europe to Central and Western Europe, and lorries also operate regularly from Western Europe into Turkey and even into Persia and Pakistan.

One of the most notable features of the motor car revolution has been the proliferation of roll on–roll off car ferries between the several islands and peninsulas.

The regional importance of water transport varies greatly: in Britain, canals and rivers since 1945 have played an ever-decreasing role (many canals have been abandoned), whereas inland waterways have never been of any importance in Iberia, Italy and Scandinavia. Nevertheless, in the Rhine basin water transport has held its own and new techniques have been developed, such as the push tug, to provide a more economical and faster service. Traffics have also changed, with a serious decline in shipments of coal and petroleum (the latter the result of pipeline development). The growth of the European Community has been an important factor in the success of the Rhine waterway and its related canals, encouraging the building of the long proposed Rhine–Danube Canal, proposals for links between the Rhine and the Maas and agitation for the Saar–Rhine Canal, despite the canalisation of the Mosel. The division of Germany and the cutting of the Elbe basin between the two major power *blocs* has led to the building of a direct all-West German route to the Elbe from the Mittelland Canal (the North–South Canal) and an East German proposal for a canal from the Elbe to the main G.D.R. port of Rostock. Growing importance of the upper Rhine to Basel encouraged Swiss proposals to make the river navigable to the Bodensee and for a canal to link the Rhine to the Rhône via Lake Geneva.

Influenced by massive Soviet inland waterway projects, a plan was conceived in Eastern Europe in the 1950s to join the river systems of the north and south together, by completion of the proposed Elbe–Oder–Danube Canal, while links between the northern rivers, particularly from the Vistula to the Dnepr, were proposed and improvement of several rivers like the Vistula and the Saale also foreseen. The building of the Iron Gates barrage on the Danube to flood the dangerous narrows has been a major first step. A project begun but later abandoned was the building of a canal across the Dobruja to shorten the Danube–Black Sea journey. Work on this vast waterway 'ring' appears to have, however, a low priority. The full potential of the Danube remains unrealised and its traffic is substantially less than that of the Rhine, though surprisingly early postwar ideas generated in America for some sort of 'TVA' scheme on the Danube have never been pursued by the Socialist *bloc* states, despite the model of the Soviet *Great Volga* scheme.

The revolution in sea transport initiated by the Suez crisis in the mid-1950s has had particularly sharp repercussions in Western Europe, where the super-tanker and the very large bulk carrier have begun to influence heavy industrial plant location. The greater international mobility offered through the European Community has acted to the advantage of very large vessels in allowing concentration of effort at suitable harbour sites, as in the case of Rotterdam-Europoort or the French plans for a megaton tanker terminal in Brittany. The offloading at Bantry Bay in Ireland of supertankers into smaller tankers for local distribution to older refinery sites is another example of this trend.

Less oriented to maritime trade, the impact in Eastern Europe has been smaller, partly because few *Comecon* ports in the Baltic or the Black Sea can take fully laden large vessels, though the Adriatic situation is better and has encouraged development plans. The *Comecon* countries have sought to develop their merchant fleets in general cargo, and the Soviet Union, Poland, and to a lesser extent the G.D.R., are now significant in the world tramp market.

The increasing quantities of petroleum and natural gas have been handled by a growing length of pipelines, for which two major patterns have emerged. In Western Europe, the orientation is markedly from oil terminal ports to inland refineries or distribution centres, but there are also internal systems (mostly for natural gas) from deposits to consumers (for example, in the Netherlands, in North Germany and from the North Sea into England). A large new pipeline system for both natural and crude oil will link sub-sea deposits to landfall terminals in the northern North Sea.

During the 1950s a switch of passenger traffic to airlines took place. Particularly marked was the decline of oceanic passenger liner traffic, with heavy losses on the Europe–North America routes, but also to Africa and the Far East. Internal air traffic built up more slowly, though by the mid-1960s it was already developing impressively. In the larger countries and on the longer intra-European routes appreciable time savings could be made. In some cases, it was the most convenient way of overcoming unpleasant frontier controls and transit difficulties, for example, between West Germany and West Berlin. Traffic is handled both by state airlines and by private hauliers (particularly the big tourist traffic). Europe has become a focal point in world air traffic patterns. In general, each country has come to have its major international airport, usually sited near the capital (the exception is the Federal German Republic), though international flights can also be made from other airports. Three-quarters of all European air traffic is handled by London, Paris, Frankfurt (Rhine–Main), Rome, Copenhagen and Schipol (Amsterdam), but one-third alone is from London Heathrow. Air freight development means that airports are increasingly significant in world trade: London Heathrow is, by value of turnover, Britain's second 'port'. There has been an attraction of industries based on air freight to sites near airports: Shannon airport in Ireland, once an important staging point in Atlantic traffic but now overflown by airliners, has become closely attached to a duty-free trading estate. Tourist traffic is often based on particular airports (for example, Luton in the United Kingdom), so that Palma Mallorca has more traffic than Zurich, Gatwick or Madrid, its closest rivals.

Peasants in the New Europe

In a setting of world concern about food supplies, the immediate postwar years over most of Europe were ones of shortage, notably in the towns, owing to disruption of distribution and marketing systems and the wartime slaughtering of animals compounded with inadequate wartime labour from military mobilisation of manpower and even the expulsion and flight of indigenous farming populations. Country people were at first in a strong position as food producers, but as industry revived or was developed, they began to see their standards of living fall increasingly behind the industrial-urban population. By the early 1950s, in Western Europe the structural weaknesses of farming were becoming more and more clearly shown – inefficiently small units, antiquated tenurial systems, undercapitalisation and gross overmanning. Mechanisation or improved techniques hardly helped and the flight from the land began to grow. By the late 1950s, marginal land and even some good land near industrial centres was going out of use, while agricultural policy created dear food and unmarketable surpluses. In Eastern Europe, application of Soviet ideas of collectivisation caused major upset and a drastic fall in food supplies, whereas priority to industrialisation and urbanisation in some countries drew too many people too quickly from the countryside (for example, in Bulgaria). By the mid-1950s a combination of economic mismanagement and peasant resistance to collectivisation was reflected in political crises, as in Hungary and Poland in 1956. The socialist *bloc* countries then began to find their own solutions instead of blind acceptance of the Soviet pattern, but the issue of collectivisation had already brought the defection of Yugoslavia from the Soviet *bloc* as early as 1948.

Agriculture has remained the most intractable of postwar Europe's economic problems, whether in the free market pattern of the West or the centrally directed planning of the East. Everywhere the share of farming in employment has fallen and the rural landscape has changed in a way not seen for at least two centuries. At the same time there has been a preoccupation everywhere with trying to equate living standards between town and country.

In both East and West, however, there has been little success in closing the widening gap between the living standards of town and country. In Eastern Europe there has been a reluctance to divert scarce investment and resources from industry to agriculture and *Comecon* has taken little interest in agrarian problems, leaving them for solution to the individual countries. In Western Europe, the Economic Community has made a multinational approach, with the creation of a Common Agricultural Policy to give farmers assured markets and realistic prices without burdening consumers with unduly costly food. Yet it has hardly achieved its aims and structurally farming still must be vastly improved: in some ways the policy has exacerbated the situation, for which the member countries pay a rising bill. Faced by escalating support costs, several drastic proposals to create a realistic farming industry have been made, though these have been so draconic that they are most unlikely to be politically acceptable in any farming region. A vicious circle of stalemate has arisen. Perhaps most significant has been the effect of the European Community's protectionism in erecting barriers that hamper trade in agricultural produce between Eastern and Western Europe and distort long-standing trading patterns with Australasia and North America.

Reference

1. J. I. Servan-Schreiber, *Le Défi Americain*, in English translation as *The American Challenge*, Penguin, Harmondsworth, Middlesex (1969).

7 Profiles

Although the ideological schism in Europe after 1945 destroyed the original hopes of a completely united approach to the continent's problems, the concept of European unity nevertheless fired many imaginative minds, particularly in the West. The organisations that emerged from these aspirations, set within the existing political–geographical framework, are slowly but inexorably changing the broader functional regional frameworks as a search for a new and hopefully realistic cohesion has been sought.

The European Community

With clearly defined aims of political and economic unity in the Treaty of Rome that empower it with truly supranational status, the European Community[1] has the greatest potential to channel and direct the dynamic and volatile forces for change developed in Western Europe. Conceived in an atmosphere of fear that Soviet influence would sweep westwards, the European Community was germinated in a belief that close association and mutual aid were the best ways to surmount the divisive and conflicting forces that had twice in a quarter of a century brought Europe to war. There was also a realisation that even the greatest European powers had been dimensionally overtaken by the new superpowers and that individually European countries could not face the new technological and political challenges beginning to arise.

Whereas all the Western European countries were agreed on the need to abandon the prewar economic nationalism and to co-operate fully in trade and technology, some had misgivings about committing themselves to formal political union. West Germany, France, the Benelux countries and Italy had suffered a greater loss of faith in established political patterns than Britain (still in the late 1940s heavily committed outside Europe), whereas in the Scandinavian countries and Switzerland there was a reluctance to political involvement on a wider front and a strong desire to retain their traditional neutral posture. In Finland and Austria, political circumstances precluded participation, while the shattering experience of the Nazi–Fascist interlude hardly made totalitarian Iberia welcome.

The Organisation for European Economic Co-operation, initiated to help effectively apply American Marshall Plan aid, had provided an opening to international co-operation by the day-to-day experience in studies of basic problems, fact-finding and co-ordinating activities among the fourteen (later

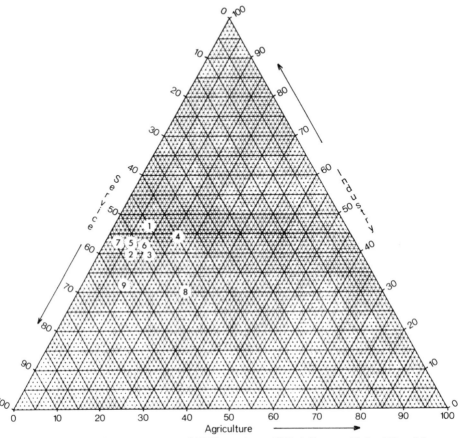

Figure 7.1 Employment sectors of EEC members, 1973. 1 German Federal Republic; 2 Netherlands; 3 France; 4 Italy; 5 Belgium; 6 Luxemburg; 7 United Kingdom; 8 Republic of Ireland; 9 Denmark

eighteen) members. There was an early appreciation of the need to draw Germany back into the community of nations without apparently surrendering the occupation statutes. A brief interlude in this process was the International Ruhr Authority (set up to regulate German heavy industry) that provided a further exercise in international co-operation. The need for co-ordinated action was also reinforced by the new dimension created by the defensive alliance of the North Atlantic Treaty Organisation that stressed a new concept of political–geographical unity in the North Atlantic basin. The major difficulty of finding a true concept for Europe and its unity has been reflected in the number of organisations – such as Western European Union and the Council of Europe – that have come close to co-ordinated action but have each time faltered at the last hurdle when confronted by the conflicts and confusions bred in a diversity of experience, spatial association and nationalism that rob European dynamism of its central direction.

The first strong functional organisation, primarily economic but containing an element of supranationalism, was the European Coal and Steel Community (1951). Designed to share out the market for coal and steel more equably and to rationalise production to satisfy the pent-up demand for these commodities, it was significant in having certain strictly limited and rigorously controlled supranational rights. It was also the ideal organisation to begin drawing Germany back into the community of nations by replacing the International Ruhr Authority and to ease Franco-German tensions over the coalmines and steelworks of the Saar. It helped the recognition of a new functional regional concept – the *Montandreieck*, the heavy industrial triangle of the Ruhr–Saarlorlux–Benelux – fundamental to the first stage in the development of a European economic community.

In the early 1950s an urgency was given to the progression from the ECSC to a broader and more integrated economic and political union by a feeling, particularly in continental Western Europe, that in a materialistic way American and Soviet interests in Europe were not dissimilar and that any confrontation or clash would be at Europe's expense. It was therefore thought desirable by many to establish Europe as a third force to avoid the consequence of any misguided 'brinkmanship' by the superpowers. The pursuance of this closer community was most actively followed by the six countries of the ECSC, resulting in the creation of the European Economic Community embodied in the Treaty of Rome (1957), aimed at an 'ever closer union among the peoples of Europe'. It was perhaps unfortunate that the rigidity of the Treaty, set in the circumstances of its time, resulted in several important potential members shying away, since they would have preferred a more flexible and slower approach to any ultimate political and economic union.

Economic and political considerations were reinforced by geographical influences in defining the original membership of the European Economic Community, just as they had been in the emergence of the ECSC, bound together by the common experience of the multinational *Montandreieck*. The new political and economic dimensions fitted well round the Rhine basin and its associated waterways and around the common territorial interests of France, Benelux and West Germany, whose willingness to come together was strengthened by their experience of the price of pre-1939 economic and political nationalism in splitting that same regional unity. Italy was attracted to the new Community by its isolation in Southern Europe between the politically unacceptable Iberia and the new Soviet imperium in the Balkans. The most important of the countries that withheld membership was Britain, which though no less affected by the war, still felt strong enough and dimensionally great enough through the Commonwealth to stand alone. Its offshore position, psychologically at least, isolated it from the common experience of the mainland countries in the period of most intense nationalism. Except to ensure that no one power dominated the continental mainland, Britain had seldom given primacy to European affairs. Britain had been preoccupied for over two centuries with the problems of a vast ocean-linked empire, and it still believed that such ties would be strong enough to maintain its world position at the centre of this motley collection of territories, currently in process of change from a centralised imperium to a commonwealth of virtually independent nations. That events had proved this otherwise had not been fully digested by the time the European Economic Community was created; harsh reality was to dawn later at a less propitious time.

Federal Republic of Germany

If we consider the individual members of the European Community, it becomes plain what contribution each has made and also what advantages as well as what problems the new regional alignments have posed. In the mid-1970s the richest state by far is the German Federal Republic, but this predominance of wealth was not so pronounced in the opening years. In the broad concept of the European movement, some of the most enthusiastic support came from the Germans, for whom a move towards European union provided a way of moving away from the occupation statutes and their constraints into a broader international community. As the West German economy revived, any opening to increase markets was welcome; but German heavy industry was equally valuable for economic growth in Western Europe and a cornerstone of the *Montandreieck*. European union to the Germans also pointed a way to removing the long-standing German–French animosity and to solving the important territorial question of the Saar's forced membership of the French economic union: it also removed outside pressures, notably the claims to German territory made by the Benelux countries.

Legislation of the European Community, freeing mobility, has had a substantial impact on West Germany. The growth in the role of the Rhine and its enhanced arterial function as well as the freeing of movement by limiting the right of states to use legislative and fiscal means to channel traffic to their own advantage have acted very much in favour of the Rhine delta ports but, on the other hand, to the detriment of the North German ports, to which traffic can no longer be artificially diverted. The ever-increasing attractiveness of a location close to the Rhine accentuated the problems of Germany's peripheral regions: North-west Germany, with its heavy refugee burden in the 1950s and lack of economic diversification, felt the trend acutely until the spin-off of the prosperity of the Rhine basin clouded some of the difference. The problem was most acute in Schleswig-Holstein, trapped between Denmark (until recently outside the European Community) and the Iron Curtain. Equally isolated from the Rhenish axis has been South-east Germany along the Czechoslovak frontier, but ties with Austria have been close, even though it is a non-member of the Community. The importance of the Rhine axis has been reflected by its steady growth in population and wealth, and along it are strung the major growth areas of the Republic. This trend generated a characteristic gradient of these two parameters from east to west, most marked in the 1960s.

The Rhine axis contains the bulk of the West German iron and steel industry, especially in the Lower Rhine–Ruhr area: indeed, the trend has been a rising concentration around the Rhine–Ruhr confluence as ore supplies have come increasingly from Rotterdam-Europoort. As energy generation shifted from coal to oil, the Rhine valley also gave a particular ease to supplies shipped inland from Rotterdam-Europoort. Although this locational advantage has remained in the period of pipeline building, this means of transport has enabled the Rhine basin to be easily penetrated also by pipelines from the North German ports, and even from the Adriatic and the Mediterranean. By the early 1970s the Rhine was marked by a chain of important refineries. There has also been development of petroleum refining and associated processes outside the Rhine basin (for example, the Ingolstadt area of Bavaria). The rapid expansion of a varied consumer durables industry from the mid-1950s was also most markedly concentrated in the Rhine basin, particularly in the Rhine–Main area and in the tributary Neckar basin.

As the effect of the European Community has been to tie German industry closely to its western neighbours through new linkages, so the significance of the Rhine basin in the economic geography of Western Europe has risen. Organisational linkages, for example, in the iron and steel industry and in heavy chemicals have drawn the Rhine–Ruhr industrial node into close association with the Rhine–Maas and Scheldt deltas. Despite its growing locational disadvantages for coal and ore supply, the Saar has used its central position within the *Montandreieck* to broaden its linkages, whose international character has been made easier by the political–geographical background of the region, as well as to strengthen its ties with the Rhine axis. The consequence has been the appearance of the trinational region of *Saarlorlux*, with close international integration in a range of industries. Similar linkages have appeared in the manufacture of consumer durables (particularly motor vehicles), with the three major Rhenish focal regions for their production still well entrenched and growing: the Lower Rhine, the Rhine–Main and the Neckar basin. These regions are excellently located centrally within the original Community of the Six market.

The powerful agglomerations of economic activity and population (with all the problems of sprawling urban and suburban settlement and heavy infrastructural investment) of the Rhine basin were made possible through the *laissez-faire* policy of the West German social market economy in the extremes of commercial freedom in the 1950s and 1960s, when planning was in the lowest possible key. After the heavy industrial crisis of the late 1950s, it became apparent in the 1960s that the regional disparity in development through the concentration of activity along the Rhine axis was producing an unacceptable steepness in the gradient of economic activity and wealth in relation to other parts of the Republic. It has nevertheless been difficult to slow the dynamic momentum achieved by the Rhine axis, though government action has sought to encourage dispersal to the north-west or south-east, reflected in major industrial developments in the Lower Elbe (for example at Stade) and in the petrochemicals triangle of southern Bavaria.

German thinking on European unity in the late 1940s had been coloured by fear of overpopulation resulting from the massive refugee inflow. Better relations with neighbours might well have provided opportunities for emigration, but the rapid revival of the economy quickly proved emigration unnecessary, even undesirable. In the 1950s refugees from the Soviet Zone (the G.D.R.) were welcome additions to the labour force, until the stream was stopped by the building of the Berlin Wall in 1961 when rapid overheating of the economy through labour shortage followed. The expedient was to let in foreign labour, mostly from southern Europe and Asia Minor outside the Community. Labour mobility from within the Community itself, despite encouraging legislation, has been low, largely for structural reasons.

The division of the country and the loss of large farming territories to Poland made the question of food supply an early postwar worry in Germany, but by the mid-1950s this fear had receded, though there was still concern that the prewar level of self-sufficiency had not been regained. There was also a dilemma of how to give the peasant the traditional protection and yet rely more on imports. The idea of a common agricultural market in Europe, where peasant protectionism was widespread, as a means of securing food was attractive. Whereas it was clear that within the Six, France, the Netherlands and Italy would gain most by secure farm produce markets, the Germans hoped that any concessions made in farming policy would be offset by the gains in a similar common market for their industrial products. Certainly, the German peasant has gained little in competition with French, Dutch and Italian producers, particularly in sectors such as horticulture, wine and poultry production. Agriculture in West Germany has been the Cinderella of the Economic Miracle. Through the 1950s the position of the peasant deteriorated and the labour demand from industry with its greater opportunities was such that many left the land. Reluctant to part with their land and yet unwilling to farm it for poor financial returns, large tracts have simply been left as 'social fallow', limiting government attempts to reorganise farming on more profitable lines.

The German lands have long served as the 'turntable of Europe', a role enhanced in West Germany through the division of the country and the creation of greater mobility in the Community. North–south traffic between non-socialist *bloc* countries in order to avoid routes across the Soviet *bloc* has been forced through West Germany. The rapid extension of the motorway system has encouraged easy movement across West German territory. Although North German ports have suffered from the easing of international movement in the Community that favoured Rotterdam-Europoort, the development of this vast complex and the growth in significance of the Rhine axis further encouraged trans-German traffic. Growth in economic activity has, however, been great enough to bring additional traffic to North German ports, despite the deflection of Rhine basin traffic from German to Benelux ports. Wilhelmshaven has emerged as an oil terminal, the Weser ports have developed as container ports, whereas Hamburg has continued as a major general port. With growing congestion of the approaches to the Channel and southern North Sea, combined with difficulties of shallow water, it is believed the Elbe estuary could offer a site for a major deep-water terminal for extremely large vessels entering from the north.

France

Much of the drive for the creation of the European Coal and Steel Community and the subsequent Economic Community came from France, the other original major member. French motivation to join the broader concept of a united Europe has been coloured by historical considerations: to eliminate the conflict with Germany and to re-establish its image and prestige in Europe and the world. France, like Germany, occupies an important nodal function in Europe: it can be considered to have three major interfaces – the Atlantic, the Mediterranean, and the Rhinelands.

Although French control of the Saarland and participation in the International Ruhr Authority were important elements in rebuilding its dislocated economy, the speedy economic revival of West Germany after 1948 demonstrated how rapidly France could again be overtaken by Germany. This feeling undoubtedly underlay the proposal by Robert Schumann to form the ECSC to 'internationalise' regulation of the iron and steel and coal industries. Better relations with Germany were important, because it was rich in energy sources, one of the critical weaknesses of the French economy in the late 1940s and early 1950s. The concept of the *Montandreieck* had a major French element, since French heavy industry was strikingly concentrated in the north and north-east, while the efforts at integration of economic structures through the Community has produced the regional idea of *Saarlorlux* encompassing a vital part of north-east France.

The switch from coal to oil and natural gas as a source of energy, although not greatly changing the French domestic energy resource position has freed industrial location and led to the development of major refining complexes on the Mediterranean coast (Étang de Berre) and Atlantic coast (Basse Seine, etc.), while natural gas is piped into the interior from the Lacq deposits and imported into northern France from the Netherlands. France will, however, remain one of the smaller West European domestic producers of natural gas and oil unless offshore deposits are found.

The effect of the integration of national economies within the Economic Community and the draw of the Rhine–Rhône axis has been to attract considerable industrial development into eastern France. The emergence of a tightly knit industrial structure extending from the Rhine–Maas delta across Belgium, Rhenish Germany, Luxemburg and eastern France (and even into the Swiss Mittelland) has overridden national boundaries to such a degree that it has been suggested that a new axial *Lotharingia* (based on the name of an ephemeral early mediaeval kingdom) has emerged.

The advantages of the Rhine–Rhône axis for specialisation, industrial interlinkages, market orientation and transport and the related infrastructures have tended to concentrate French development primarily in a broad angular crescentic axial belt across northern France, from the Channel coast through the northern coalfield and Lorraine orefield to the Rhine and into the Rhône basin. Le Havre and Dunkirk have established bulk carrier facilities, the first for oil and the latter for iron ore and coking coal, whereas the ferry ports have become more significant as the trade with Britain has developed. Nearness to this axis also accentuated the commercial and financial importance of Paris. The Rhine–Rhône axis extends south to the Mediterranean coast, where petroleum refining and petrochemicals have expanded. Outside these axial belts and the Paris region has lain the area dramatically described as the 'désert français'.

To counter this spatial imbalance – seen similarly in West Germany – the French Government has encouraged developments in Brittany, south-west France and the Massif Central. Having accepted planning earlier, with the advantages of direction offered by central government, they have been a little more successful than the West Germans, where federal particularism has adversely influenced spatial planning at the national scale. The French would have particularly welcomed the Channel Tunnel to encourage development in the rather isolated Nord and Pas de Calais, whereas Britain was afraid the tunnel would have encouraged further drift towards saturation in South-east England.

The continuing importance of agriculture in the French economy, and especially the large and varied surpluses of foodstuffs available for export, put France in a special position to benefit within a large assured common agricultural market. Like Germany, the main difficulties have been structural – the need to modernise the undercapitalised small peasant farmers, many of whom still farm on a quasi-subsistence basis, while the rapid growth of industry has widened the gap in wealth between town and country. But, in general, both the larger size of holdings and their greater natural productivity put farming in a stronger position than in Germany and endowed France with greater self-sufficiency in foodstuffs.

The increased circulation brought by the easing of mobility has affected France less in the Community than the more centrally placed Germany. The long but relatively lightly loaded French railway system has felt the main growth of traffic in the north, north-east and along the Rhône axial extension from the new Rhenish focus, though Paris has remained an important centre of international passenger traffic. Whereas the national energy problem has encouraged railway electrification, this has also been most marked in the north and east. The motorway system radiating from Paris has developed less than in Germany or Benelux. Links to the north and east are still poor, though the major Paris–Marseilles axis is complete. Particularly impressive has been the growth of Le Havre and Marseilles, primarily as oil-importing ports, despite not being well placed to penetrate the Rhine axis by road, rail or waterway. Far less growth has been shown by the south-western ports (Bordeaux, Nantes) or even Cherbourg. France is poorly linked to the Rhine waterways, though possible Rhine–Rhône links have been discussed: the bulk of present waterway traffic is on the Seine below Paris and on the Rhône. Paris is the major airport focus of the European mainland, though in total traffic considerably behind the London airports.

Adjustment to the new Community has gone along with France's decline as an imperial power. Particularly sensitive were the losses in North Africa, where Algeria had been regarded through its many French colonists as a virtual part of Metropolitan France. Whereas in the interwar years French links with the Rhine basin had been undermined by German hostility, the *rapprochement* of the de Gaulle–Adenauer era gave an impetus to France to turn towards the Rhine basin.

The Benelux Countries

The Benelux countries – Belgium, the Netherlands and the small but economically important Luxemburg – have become a focal region in Western Europe: they form the core of the striking north-west European 'megalopolis' and the heart of the great 'Eurodelta' of the Rhine, Maas and Schelde, increasingly dominated through the 1960s by the immense expanding port facilities of Rotterdam-Europoort.

The Benelux countries were in the vanguard of the movement towards European unity, since these small states had no doubts as to how much their destiny depended on the most cordial relations with and between their larger neighbours – for the Benelux countries had suffered harshly through the clash between these powerful states. Nevertheless, there were also some reservations, because they were anxious to retain their own freedom of action rather than become dominated by larger powers, particularly if France sought to replace a broad Community approach by a rigid Franco-German alliance. Against such an eventuality, Benelux members, especially the maritime Dutch, were consequently anxious to maintain and develop their traditional relationships with Britain.

Although small, all three are in relation to their size important industrial countries. Luxemburg is distinguished by its heavy iron and steel industry in the south, in structure and *raison d'être* similar to the adjacent industries of the Saar and Lorraine. Through the EEC, Luxemburg has benefited by diversification of its monolithic industry owing to the penetration of firms from other Community countries bringing in tyre manufacture, plastics and light engineering, while it also gains from the community of interests established in the concept of *Saarlorlux*. Belgium also has its own iron and steel industry, linked likewise with the Luxemburg and Lorraine industry, though mostly sited along the Meuse valley (the old Wallonian coalfield) or further north-west, where capacity has grown near the coast, as in the modern plant in Gent. Belgian coalfields have undergone more drastic rationalisation than in any other Community member: the older fields, notably in the Borinage, have been abandoned in favour of richer deposits in the northern Kempenland. Imports of natural gas from the Netherlands have become an important energy source, and crude oil and products are imported from Rotterdam. The southern Belgian Walloon industrial districts have been overtaken in significance by the newer north–south industrial axis between Brussels and Antwerp: to counter this challenge, the southern iron and steel industry (for example around Charleroi) has become increasingly specialised. The Flanders textile industry has gained a wider market, despite the common Community problem of cheap imports from Eastern Europe or the Far East. Light engineering has grown by making parts and assemblies in business linkages with other Community countries. With a major ferry traffic to Britain, Belgium is an important transit land, while Antwerp has flourished as the old constraints of its position have eased through better relations with the Netherlands and much spin-off traffic from Rotterdam-Europoort. The consequence has been formidable industrial investment in and around Antwerp (much German and American capital). Its position will be further strengthened as better inland waterway links to the Rhine–Maas are completed in the *Delta Plan*.

Economic trends have, however, caused a major shift in Belgium's ethnic problem. The more rapid growth of population in the Flemish north has been strengthened by its rising economic wealth and the Walloons have lost their earlier dominant economic and social position. There have also been growing pressures for federalism, reflecting perhaps a new feeling of security within the Community.

Controlling the Rhine–Maas delta, the Netherlands is a commercial and maritime nation closely bound to the Rhine basin, though not always happily in the past. The value of the Rhine trade to Rotterdam and other Dutch ports made early association with any common market in the Rhine basin highly attractive, especially as the freeing of mobility could only work to the Netherlands' advantage. However, reservations included a fear that such a small country would be dominated by its larger neighbour and that, as a maritime country, its interests might be ignored in a predominantly continentally oriented Community, reflected in the underlying Dutch support for the closer association of Britain in the new experiment.

The core of the Netherlands has become the immense annular conurbation of the *Randstad Holland* and the world's largest port, Rotterdam-Europoort, whose huge concentration of oil refineries and special quays are linked by pipeline, waterway and railway to the German Rhenish industrial areas. It has gained over Antwerp by its easier access for very large vessels and easier routeways into the interior. The pattern of railway electrification, motorway construction and waterway improvement in the Netherlands has illustrated clearly how tightly bound the country has become with the Rhenish interior, while as an international airport, Amsterdam-Schipol closely rivals Frankfurt-Rhine Main. The main problem for the future is how much more traffic by very large vessels Europoort can handle and whether this can be effectively regulated in the narrow Channel approaches, though a large potential remains to develop in ferry traffic with the industrial English Midlands and North via Humberside.

Immediately after the Second World War, when much coal was imported, considerable investment was made in the coalfield of Dutch Limburg, but as these deposits would not suffice long term, claims were made for German territory in the hope of extending the coal reserves. The situation changed

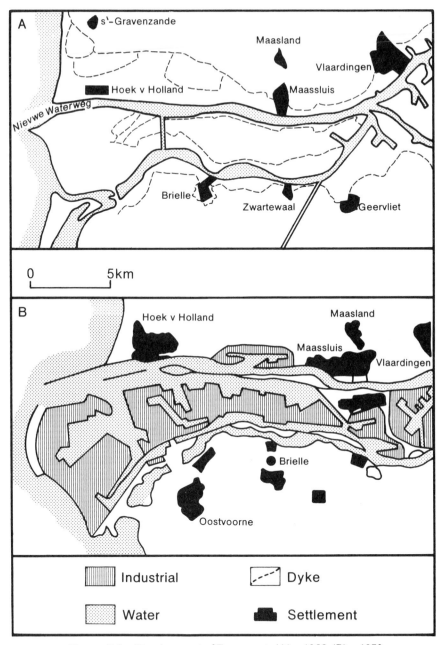

Figure 7.2   Development of Europoort: (A) *c.* 1955; (B) *c.* 1973

rapidly after 1959, with increasing discoveries of natural gas, especially the large deposits in the Groningen province. These have made the Netherlands not only self-sufficient but also allow a large export. This source of energy so overwhelmed the domestic market that some modern and expensive investments in the coalfield have already been closed.

Although less important as an iron and steel producer than Belgium, the Netherlands has the advantage of potentially good deep-water sites for modern bulk-carrier based iron and steelworks. Already the Ijmuiden works on the North Sea Canal have linked with German works in Dortmund to use such advantage. Plans have been formulated for a large complex at Europoort, though formidable labour, infrastructure and environmental problems beset the site. Otherwise, manufacturing industries are scattered in the *Randstad* and especially in Brabant and Limburg. The large electrical engineering industry of Eindhoven, seat of the multinational Philips Company, is particularly important, and the town also has the DAF motor vehicle plant. Textiles are the main industry in the north-east Netherlands, though natural gas deposits have attracted other industry. The Rhine–Maas delta has witnessed much foreign investment in container handling terminals, chemicals plant and refineries, while Rotterdam's shipbuilding remains of European significance.

The Benelux countries, though small and densely settled, have important agricultural sectors in their economies, especially the Netherlands. Horticulture and poultry rearing produce significant export surpluses, but the Dutch are also major exporters of dairy produce – butter and cheese – and there are large related surpluses of veal, while foodstuffs imported and processed around the Netherlands' ports are also re-exported, notably to the British and German markets.

Italy

The sixth original member of the Community, Italy is poorly endowed with mineral resources for heavy industry and energy, but with substantial manufacturing capacity and labour it has been in a special position. Earlier this century, its vigorous population growth had been kept within the bounds of home resources by a steady emigration, either to its mediocre colonial lands or overseas, notably the Americas. Not a rich industrial country, Italy had been badly damaged by the long slow invasion of 1943 and deprived of its former colonies: plans in any form to aid European reconstruction were therefore welcomed. A broader European Community offered not only markets for Italian goods and agricultural produce, but also a source of vital raw materials in advantageous trading conditions, as well as a sponge for surplus labour.

In the late 1940s Italy found itself in an isolated position in the Mediterranean. The southern shore of the sea was increasingly involved in an Islamic-nationalist struggle that contained a strong anti-Italian element. On the north shore, relations with Greece and Turkey were poor and contact with Iberia was poisoned by the peninsula's totalitarianism. The Adriatic was dominated by hostile Yugoslavia and Albania, both in the socialist camp. By the nature of the physiography of Europe, Italy has many characteristics of an island state, with sea-borne contacts with the world at large easier than the difficult transalpine routes, despite the construction of road and rail tunnels in the modern period.

Like much of southern Europe, Italy's relative position in Europe had slipped back during the nineteenth and early twentieth century coal and iron age. The economic progress made since the Second World War has therefore been all the more remarkable, with the main spurt coming since the inception of the Treaty of Rome in 1957. The annual rate of growth of industrial production has been among the highest in the Community (though less than in Spain), while the rate of growth of the industrial labour force has been the highest in Western Europe. Italy has supplied large numbers of workers to Germany, France and Switzerland.

The change in energy patterns has helped Italy, even though its own petroleum resources are small, but natural gas is more generally available. Extensive areas of the country are, however, potentially petroliferous. Italy has developed as a major oil refiner, using imported African, Middle Eastern and Soviet crude oil, and Genoa and Trieste have emerged as significant pipeline terminals pumping crude oil into Central Europe. Italy also plans to import natural gas via a pipeline from the Soviet Union. Refining has been developed at deep-water anchorages in the south, where products are distributed by sea. The energy problem also stimulated use of less conventional sources, such as geothermal power, and *Euratom*'s major research centre is at Ispra. Much has been done to harness hydroelectric potential, both in the Alps and the Apennines, where the schemes are often part of land amelioration, drainage and water supply projects.

The large volumes of low cost labour available from the rurally overpopulated south of the country have been reflected in the development in the north of labour-intensive assembly industries, like consumer durables and motor vehicles (Turin is the largest single centre for vehicle manufacture in Europe). Historical factors have generated these industries principally in the northern triangle of Genoa–Milan–Turin, and two-thirds of all industrial employment is in the north, which has shown vigorous population growth. Labour was also a significant factor in the survival of a traditional textile industry, again markedly represented in the north, but most important in total employment in the centre and south. Much of Italian success can probably be attributed to the ready access to the immense and appropriate Community market.

Bulk carrier movements of coal and iron ore in the development of waterside iron and steelworks have had particular reference to Italy, where a substantial coastal iron and steel industry has developed (Genoa, Piombino, Naples and Taranto). The southern plants are designed to bring employment and diversification to a problem region. Inland plants, mostly in the north, concentrate on electric steel from scrap. Aluminium is also important at the head of the Adriatic, though its original ore supplies in Istria have been lost to Yugoslavia.

Although all the original six Community members have regional problems, none is perhaps more acute than in Italy. The south, long neglected, has suffered from archaic and inappropriate social and economic organisation.

South of Rome, rural poverty, lack of opportunity and an inferior infrastructure contrast to the obviously more affluent north, pulsating with expanding industry in the 1960s. The *Cassa per il Mezzogiorno* founded in 1950 was a radical step to co-ordinate central government funds for development in the south. It was designed for long-term planning and co-ordination of economic projects, aided after 1957 by a special Committee of Ministers for the South. The need has been not only to provide public works to better the regional infrastructure, but also to overcome long-standing problems of agriculture and tenure. Much has been done for direct industrialisation and all public corporations have been obliged to put 60 per cent of their investment in the south.

Of the original six Community members, employment in farming is highest in Italy and agriculture remains a vital sector of its economy. The Community has provided a useful market for many Italian products not well represented in other member countries, notably citrus and other fruits, rice, vegetable oil and distinctive wines. The focus of commercial farming is, however, the north, directly in competition with industry, whereas inefficient, poor yield, extensive farming, apart from a few favoured and irrigated areas, is characteristic of the south.

Italy has a peripheral location in the Community (most acute in the problematical south) and its main links are the transalpine road and rail routes, most of which necessitate crossing non-Community territory, though existing arrangements for such traffic raise few problems for a substantial growth in volume. Though influenced by the decline of the Suez Canal as a link between Europe and the East, Italy has been fortunate in the number of deep-water ports able to accommodate bulk carriers, and the country has taken its share of growing Mediterranean movements of crude oil and liquid petroleum gas as North African oilfields have developed, especially as its northern ports are better located as pipeline terminals for Central European markets than ports in southern France.

In 1973 the Community was enlarged by three additional members – the United Kingdom, Ireland and the sole Scandinavian member, Denmark. Sympathetic to the European idea from the outset, several factors had restrained these countries from membership of the original Treaty. The largest new member, the United Kingdom, is likely to considerably change the Community and to tip delicate balances established between the original six members, especially the political–geographical implications for the small members. Apart from the additional population and industrial potential, the United Kingdom adds a significant spatial component. The main 'Euro-axis' of the Rhine is extended across the southern North Sea by inclusion of the English axial belt from the Mersey basin via the Trent basin to the Thames. As an offshore island, United Kingdom membership turns the North Sea into a virtual Community Lake, but the true role cannot be achieved until really effective transport links, either as a tunnel or high capacity ferries, to the European mainland have been established. In some respects, Britain has similar problems of peripherality in the Community that beset Italy, though the former – at least in its southern areas – is better placed nearer the Community core than Italy.

The United Kingdom

Many supporters of the original Community felt that its long-term aims were stunted without Britain, though British application for membership was not without its difficulties after the important psychological moment at inception had been lost. Suspicion of long-term Franco-German aims certainly gave support for Britain in Benelux and in Italy, while opinion in Germany also felt a need for Britain's participation. Most opposition came from France – particularly from the gaullists – where Britain was frequently regarded as 'of Europe but not in Europe'. The root of French objections was a traditional belief that Britain was set to thwart French aspirations of European leadership. Unfortunately, many actions and views in Britain did little to alleviate fears of the French school.

The supranationalism of the Community, on the other hand, was a British fear, particularly if it were to reduce sovereignty and restrict ability to manoeuvre in foreign policy. In the early 1950s the much broader horizon of British foreign policy contrasted with the 'local' dimensions of the Continent, so the misgivings were not irrelevant. Britain, like the rest of Europe, was, however, keenly interested in a Franco-German *rapprochement*.

Wise after the event, a primary failing was not so much an underestimation of Europe's future role in British affairs as an overestimation of the long-term significance of the Commonwealth to Britain. This has been exacerbated as the Commonwealth has become strongly identified with Afro-Asian interests and the ties with the white ex-Dominions (S. Africa, Australia, New Zealand, Canada) have loosened. There has also been the vague and illusory 'special relationship' with the United States that made many in Britain reluctant to commit themselves to a true European image. In any case, the United States shifted increasingly in favour of the Community. Continental aspirations were clearly towards federalism and a more defined international structure in economic relationships: Britain favoured free trade and intergovernmental co-operation based on goodwill and a minimum of strings. Britain was reluctant to alter Commonwealth preference, with cheap food and raw materials, for intangible and as yet uncertain advantages, whereas the Continental countries were not prepared to expand their horizons to encompass more than Europe. In hindsight, both groups missed golden opportunities.

Once the European Economic Community got under way, it became increasingly clear that Britain could not afford to be on the wrong side of the tariff barriers for this immense market, however problematical equating membership with existing Commonwealth relationships might be. The British defeat in the Suez crisis and the growing independence movements in many Commonwealth countries in the late 1950s made it increasingly clear that British resources were simply not enough to go it alone in the high technology that the advanced industrial countries were to rely on to a greater extent, as industrialisation in the less advanced countries closed oncelucrative markets. Abortive attempts to gain membership have been described elsewhere: only in 1973 was full accession achieved.

Coalmining, the basis of the original British industrial structure, has become increasingly concentrated in the eastern Pennine flanks and adjacent lowlands, and though significant fields remain in Scotland and South Wales, this concentration is likely to increase (for example, new seams opened near Selby). Britain has taken the lead as bituminous coal producer in the Community, though little of its once great export trade to Europe remains. The generous coastline of Britain, with its many deep-water harbours, offers opportunities to take full advantage of modern waterside locations for massive iron and steelworks using bulk carrier deliveries of raw materials. Nevertheless, as in West Germany, home supplies of good-quality coking coals remain important, despite their relatively high cost. Adjustment to the new trends of large integrated waterside steelworks will, in the long term, account for the overwhelming part of British capacity being sited in South Wales, North-east England and the Clyde estuary, though Scunthorpe is near enough to deep water to switch from low grade home ore to high grade imports. As elsewhere in the Community, inland works are becoming specialist producers.

Community tariff walls may help to protect nineteenth-century industries now in difficulty, such as textiles which have suffered from cheap Asian imports, difficult to resist under the older Commonwealth preference scheme. High-class woollens have a good future in the Community market as well as some branches of cotton textiles, such as spinning, since yarn is exported to the Continent. The situation of the British textile industry is not unusual – the German industry suffered in the late 1960s from East European competition inside the Community market. Although British shipbuilding suffered in the 1950s from competition from rebuilt West European yards, these yards are now also facing acute competition from yards in *Comecon*, Japan and even South Korea, raising a demand for Community counteraction. Britain has turned to building oil rigs and platforms, with major competition in Europe coming from Norway.

With good coastal sites for bulk imports, adequate home resources of coal and an expansive refinery industry, Britain has added an important chemicals industry to the Community, falling immediately behind Germany. ICI ranks about fourth producer in Western Europe, though it has only half the turnover of the three largest German firms. Coastal sites give particular importance to the Mersey, the Severn estuary, South Wales and North-east England, while refineries on the Lower Thames and Southampton Water add further sites, but local salts account for this industry in Cheshire and Teesside and the coalfields remain important.

The United Kingdom since the Second World War has witnessed the same upsurge in consumer durables seen in other Community countries. The development has been particularly in the English Midlands and the South-east, but these assembly industries have been used as an instrument of regional policy and spread with government backing. Motor vehicle building has been established in Merseyside and in Scotland, whereas domestic appliance factories have opened in South Wales, and encouragement given to firms to move to Ulster and North-east England. Many firms are closely tied to related plants on the continental mainland and the American multinationals are well represented.

Discoveries of natural gas in the southern North Sea triggered off a wider search that has led, since the early 1970s, to considerable reserves of petroleum being located in the northern basin in the British and Norwegian sectors. Even if modest by Middle Eastern standards, these discoveries recast the position of Britain, which could come close to self-sufficiency before the end of the century. Britain has been put in a position that demands a high technology support industry and services to make exploitation and further exploration possible, not unlike the position of the early 1950s when the British nuclear power industry was well ahead of Europe, though Britain still generates more of its electricity from this source than any other Community member.

The enormous research and development costs of high technology call increasingly for a truly European effort. This is a field in which Britain can make a major contribution, demonstrated in the Anglo-French *Concorde* project. Had public attitudes and economic conditions remained the same throughout development of the project, this could have been a high technology success story and encouraged wider commitments of this kind. Indeed, the collapse of the British and French aerospace industry would leave Western Europe almost entirely dependent on America or even Russia.

Although the internal transport system of the British Isles is well developed, there are problems in linking it to the European mainland. For example, though railways have the same track gauge in Britain and Western Europe, the smaller British loading gauge (a product of railway history) complicates interchange of rolling stock, but the broad gauge Irish railways make the difficulties even greater. The British motorway system, a late starter, has a radial pattern from London to the north and west. It is surprising, in view of the heavy British dependence on overseas trade, that the system is poorly linked to the major ports, particularly the gateways to mainland Europe, and better motorway links to the principal ferry ports seem an undeniable priority. The ferry ports of the south and east coasts have demonstrated by their growth the importance of adequate links to the European mainland. The Channel ports are hampered, however, by their ferry traffic running transverse to the movement of large sea-going vessels. The Humber ports give good traffic access from the Rhineland to the North of England, whereas the good position of Harwich and Felixstowe for Rhineland traffic to the Thames basin has also been reflected in their growth. Development of ferry traffic to ports in North-east England and eastern Scotland has been disappointing, though it was beginning to generate by the mid-1970s.

British west coast ports, well placed to catch and distribute Atlantic traffic, include several capable of development for the largest vessels, but lack quick and easy communication to the European mainland in comparison with, say, Rotterdam-Europoort. Perhaps two important chances have been lost to catch a greater share of Community traffic: the abandonment of the Maplin project and the cancellation of the Channel Tunnel. An adequate tunnel could have revolutionised British contacts with Europe. Nevertheless, much of the conceptual planning appeared too short-sighted, with consideration of the tunnel primarily as a gateway to South-east England and the London area, rather than as a direct access to the industrial Midlands and North of England. The elimination of much of British peripherality in the Community would seem to lie in the building of one or more Channel tunnels or their

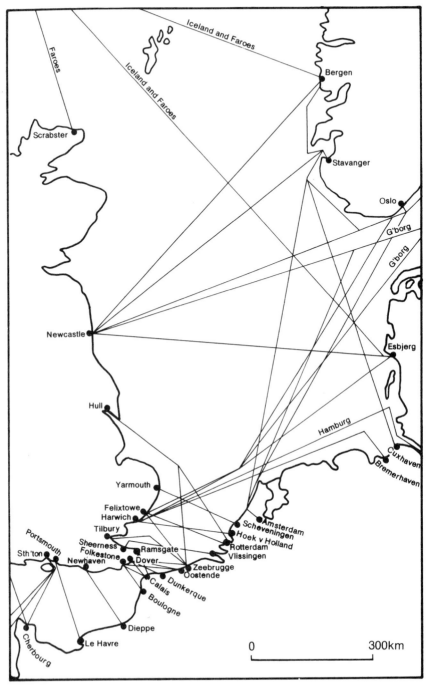

Figure 7.3 English Channel and North Sea ferry services, 1978

equivalent to deal with a volume of traffic that conventional sea ferries could not entertain.

Employment in farming is lower in the United Kingdom than in any other Community country, while the nature of farming is also different. British farms are remarkably large and compact compared with most Community countries and their economic performance compares favourably with labour intensive peasant farming. In a properly rationalised Community agriculture, British farmers would seem to have a major role as suppliers of meat and dairy produce, with a possible considerable stimulus for the widespread British hill-farming. British farming has been more successful in 'shaking out' overmanning and in applying mechanisation effectively in large fields: important economic operations not yet achieved on any scale in mainland Europe.

During the interwar years, the older heavy industrial and mining areas of the North began to lose their importance to new industrial areas (with a strong consumer durable element) and high technology industries in the more attractive environment of the English Midlands and the South-east, a trend accelerated after the Second World War by the growth of the tertiary sector. Various forms of government aid have been tried to bring new life to the older areas, but population and economic life have become increasingly concentrated in a rectangular area, with its main axis from South Lancashire and the West Riding of Yorkshire to the Lower Thames, and with growth particularly fast in the South-east in the *laissez-faire* planning of the 1960s. Nevertheless, with the development of North Sea petroleum some revitalisation of the northern areas may be expected.

The Republic of Ireland

The accession of the Republic of Ireland, though less significant, has not been without change in established Community patterns. Ireland is an essentially agricultural country, with a strong livestock and dairy basis, closely involved with the United Kingdom in economic terms. Accession has enlarged its markets, though its agriculture needs modernisation, and it has considerable potential labour reserves so that numerous Community firms already have industrial branches in the Republic. The most notable development has been the Shannon Airport industrial estate, designed particularly to penetrate the American market. It has also taken advantage of deep-water anchorages in its south-western sea-lochs to establish a major oil terminal at Bantry Bay, where supertankers can offload into smaller ships. So far, the European Community, despite its protestations of political unity as a long-term aim, has done little to help solve the long-standing political grievances over Irish unity.

Denmark

Denmark is nowadays effectively an industrial country with a major and successful agricultural sector. The dependence of the latter on Western European markets was a prime factor in the country's decision to enter the Community, particularly because of Denmark's role in the British food market.

Denmark's accession poses a regional problem for West Germany, since it is now felt increasingly necessary to bring up the level of farming and its rewards in Schleswig-Holstein to that of adjacent Danish territory. Denmark also provides a Community foothold in Scandinavia and a stronger presence in the Baltic, while it is closely related to developments in the North Sea Space (chapter 8).

Relics of the European Free Trade Area

Some countries that felt unable to join the European Community neverthe- less felt a need for closer international co-operation, though more through free trade and intergovernmental co-operation. These seven states came to- gether in the European Free Trade Association (EFTA), a loose organisation with the minimum of regulations formed exactly for the purpose of its title. It also reflected a continuance of an unsuccessful attempt among OEEC mem- bers to form a wider trading community. EFTA was considered as an interim measure, a waiting room, since it was believed that Western Europe would not long remain divided, so within a short time a much wider Community would emerge, encompassing the EFTA states.

Of the seven EFTA members, the United Kingdom was the most powerful and influential; but also significant were the three Scandinavian countries – Norway, Sweden and Denmark. Special political–geographical circum- stances were operative in the two Central European members, Austria and Switzerland, while Portugal was also a member. Later Iceland joined and Fin- land (tied by a rigorous treaty with the Soviet Union) became an associate. Even before the United Kingdom and Denmark left on accession to the Com- munity, it is hard to put a geographical dimension to EFTA. It was perhaps most heavily weighted towards Scandinavia, where it served to provide an integrational element among the Scandinavian countries, which the limited Nordic Council agreement had not sought to create. In the late 1960s a formal element of economic co-operation envisaged through *Nordek* failed to ma- terialise. One of the most clearly defined cultural, political and economic re- gions in Europe, it is perhaps surprising that the Scandinavian world has not played a more distinctive role in European integration, especially in view of its internal achievements in social and employment policies. A basic polit- ical–geographical problem is that, although fundamentally in sympathy with Western Europe, every move has to be seen against possible Soviet reaction.

Sweden

Sweden, the most important Scandinavian country, has always followed a neutral policy that made involvement in the Community unattractive, despite the close economic ties with Western Europe. Large and sparsely populated, Sweden has risen since the mid-nineteenth century from being one of the poorest to one of the richest countries in Europe. Its ores, metals, engineering goods, wood, pulp and paper have found Western Europe to be a major market, but its market in Central Europe was initially gravely dis- rupted by the creation of the Soviet closed economic system after 1945. The rich Western European market has become especially attractive for the high quality Swedish engineering industry, while Swedish ball and roller bearings have been vital to the Western European economy.

Norway

Norway depends heavily on fishing and shipping, though Western Europe is a major purchaser of these activities. It has also developed its rich hydroelectric resources for electrometallurgy and electrochemicals, again looking to Western Europe as a market for their products. Norway's position has changed radically since the late 1960s, owing to the discovery of rich oil and gas deposits in its sector of the North Sea, but because of the submarine topography the fields will pipe most oil and gas to destinations outside Norway (for example, oil to Teesport in North-east England, gas to Emden in West Germany). Sea-loading techniques for tankers from moored buoys have been developed to enable some oil to be landed directly on Norwegian territory. The changing role of coal has made the Norwegian coalmines on Spitzbergen less attractive, but strategic considerations and prospects of oil and other developments maintain a strong long-term Norwegian interest.

Finland

Finland has been tightly tied to the goodwill of the Soviet Union through its geographical position and by a rigorous treaty, which has conditioned its relations to the rest of Europe. It has had to overcome serious problems created by refugees from territory annexed by the U.S.S.R., requiring, for example, subdivision of farm holdings and the settlement of inhospitable northern areas. Although wood-using industry and metal mining (copper, nickel, titanium and vanadium) remain significant, it has also developed high quality consumer goods industries (textiles, glass, food), and a shipbuilding industry concentrating on special vessels (for example, icebreakers).

Iceland

Iceland, a latecomer to EFTA far out in the Atlantic, is generally regarded as part of Europe, to which its cultural and economic ties gravitate. Small and with few natural resources, it depends primarily on fishing. Iceland's attempts to conserve fish stocks have encouraged its efforts to extend its control over ocean waters up to 200 miles from the coast and have brought conflict with other countries, notably Britain and West Germany, precipitating extension of the Community limits early in 1977.

Austria, Switzerland and Portugal

Austria and Switzerland, Central European members of EFTA, have very close economic ties with adjacent Community members, but in the former the peace treaty precludes Community membership and the latter's traditional policy of non-involvement made such membership unattractive. Switzerland occupies a key position in the transport scene in Western and Central Europe, commanding the major Alpine tunnels, whereas Austria is also a significant transit land between Western Europe and Danubia. Both countries have used their physical environment to harness hydroelectric power and electricity plays a vital role in their economies, even to the degree of exporting current to neighbours. In both countries, electricity underlies a high quality electro-metallurgical industry. Switzerland has established a world reputation for precision engineering, notably watches and timing mechanisms, but it is also known for pharmaceuticals and textile machinery. It depends completely on raw materials from outside, especially from Community members. Austria has a well-developed if modest engineering industry, but is better endowed with natural resources: important magnesium deposits occur in the south and at Erzberg is Europe's largest iron ore deposit, supplying works at Donawitz and Leoben. Both countries depend heavily on tourism, and Switzerland is the seat of several international agencies as well as being a banking and financial centre, perhaps its most vital contribution to the world economy.

One of the poorest EFTA members has been Portugal, burdened rather than blessed until the early 1970s with a large colonial empire. It remains essentially agricultural, with 40 per cent of employment in farming. It is a major supplier of wine, cork and cellulose as well as fish (sardines and anchovies) to the European market, but industry has developed on its plentiful and cheap labour (notably textiles and domestic goods) through foreign investment.

The Uncommitted States of Non-Socialist Europe

Spain remained outside the wider European organisations until the mid-1970s, owing to the ostracism of the fascist Franco regime. Basically a poor country, it took over a quarter of a century to recover from the devastation of the Civil War (1936–1939), whereas the position was made worse because of the exhaustion of several of its once-valuable mineral resources. The large reserves of workers made Spain attractive, from the 1960s, as a country in which to develop labour-intensive assembly industries: several multi-national corporations have considered putting further European investment into Spain (or Portugal) rather than high cost, scarce labour areas in the more industrialised countries. Industries of the heavier type are mostly in northern Spain, where iron mining was important last century, and both the separatist areas of the Basques and the Catalans enjoy considerable economic significance. In the late 1970s Spain, now under a more liberal regime, has been considered a potential additional member of the European Treaty of Rome.

Greece nearly fell to the Soviet *bloc* in the immediate postwar period, but it has recently sought association with the Community, on which it depends to sell its agricultural produce, to market its shipping capacity and from which to attract tourists. Strenuous efforts have been made to modernise and increase agricultural productivity, but its small industries are chiefly limited to the main towns. Some bauxite and chrome are mined. Greek shipowners control about one-third of the world merchant fleet, including a considerable proportion of the large tankers.

Possible admission of Spain, Portugal and Greece to the Economic Community raises innumerable problems, owing to the nature of their economies, but a strong lobby sees this as less important than the defence of their fragile democracy within the Community.

Newly independent states include Malta and Cyprus, both dependent on tourist flows from mainland Europe, though Cyprus mines copper, asbestos and pyrites, whereas Malta has been left with large shipyards after the withdrawal of the British fleet. Of the micro-states, Liechtenstein sandwiched between Switzerland and Austria has been peculiarly successful in attracting light industry and is now one of the wealthiest and most industrialised countries in Europe in relation to its size.

Comecon and Eastern Europe

Soviet military and political success in the Second World War brought it domination of East Central and South-eastern Europe. The acceptance of Marxist–Leninist dogmata in social and economic organisation by the nation states of this part of Europe has led us to regard them collectively, so that Eastern Europe, as now applied, is a spatial extension of one of Europe's macro-regions within a framework of changing regional concepts conditioned by shifts of political geography.[2]

The new Eastern Europe has been emphasised by its isolation from Western Europe through the Cold War of the 1950s and by Soviet inability to accept American Marshall Plan rehabilitation aid, which Soviet satellites (particularly Poland and Czechoslovakia) were likewise forced to reject, despite their desperate need. The alternative, offered in 1949, was the Soviet-sponsored Council for Mutual Economic Assistance (*Comecon*), of which all the European socialist *bloc* countries except dissident Yugoslavia became members. Through the close integration of the member countries' economies, the Eastern European nation-states have become closely tied to the Soviet economy (at least 40 per cent of their trade is generally with the U.S.S.R.). It has been accepted that common application of Marxist–Leninist dogmata provides the political unity, though *Comecon* has stressed the national sovereignty of its members in contrast to the European Community's supranationalism.

Nevertheless, *Comecon* has had a considerable impact on the spatial pattern of Eastern Europe, which has turned eastwards to links with the Soviet Union, whose political influence, resource endowment and economic potential completely overshadow the other members, in contrast to the more even balance of these parameters between the members of the European Community. Although the supranationalism of the European Community has encouraged international trade and interchange as well as labour mobility, these elements in spatial change have received little attention or even active discouragement in *Comecon*. Whereas each national economy has developed freely within the broad interdependency of the European Community, initial policy in *Comecon*, following Soviet experience, was a search for a near impossible self-sufficiency in each member country that served only to emphasise the links with the overwhelming industrial and resource strength of the U.S.S.R. When this policy displayed its weakness and was also substantially changed in the Soviet Union, a new idea of the socialist international division of labour, involving countries in specialisation in what they could do best and consequently greater trade by exchange, was tried, with subsequent spatial change. Only from the early 1970s has a new concept of socialist integration – the close linking of national plans and the encouragement of multinational enterprises – brought some element of supranationalism, the spatial effects of which could be great but have not yet (1978) begun to show very strikingly.

While it would be idle to deny the success of *Comecon*, this has not been achieved without strains, hardly surprising with such a disparate group of members. In its early years it was overwhelmingly dominated by the Soviet Union, which remains disproportionately influential. The tumultuous events of 1956 in Poland and Hungary were the first serious culmination of tensions through attempts to get some liberalisation of the rigorous and often almost narcissistic Soviet policies: subsequent pressures by the Eastern European members for more sophisticated trading and financial structures spilled over in the Czech crisis of 1968. At the same time, emergence of differing postures among the East European members themselves have generated other strains; for example, attempts by the more developed economies to force Rumania to remain a supplier of raw and semi-processed materials rather than to develop its economy to a degree that it might become a serious competitor to the present industrial leaders. At a lesser level, competition for Baltic traffic through their seaports has brought ill-feeling between Poland and the German Democratic Republic. To try to make up lost ground during the period of forced acceptance of Soviet closed-system policies generated in the 1930s and carried over little changed into the 1950s, *Comecon* members have been increasingly buying Western know-how and even seeking Western joint investments. It may well be the end of the present century before the price of self-sufficiency policies in every aspect of economic and political life ruthlessly enforced in the 1950s and well into the 1960s has been paid.

Unlike the European Community, where full membership is restricted to European countries, *Comecon* will accept any country prepared to cooperate in its socialist (Marxist) economics. Under the sponsorship of the Soviet Union, the European members fall into a group of more advanced and sophisticated economies in the north-west – the G.D.R., Poland and Czechoslovakia – to which on several counts Hungary may now also be attached, and a south-east, more truly Balkan, group of relatively unsophisticated and developing economies – Rumania and Bulgaria– to which the dissident Yugoslavia and Albania (since 1961 no longer an active *Comecon* member) may also be conveniently grouped.

Poland

Poland's main postwar problems have arisen from the drastic change in the spatial pattern of the country as a result of the great territorial shift experienced by annexation of well-developed German territory in the west and the loss of a vast if poor territory to the Soviet Union in the east. It had also experienced one of the greatest losses of population among the Eastern European countries, owing to a combination of war deaths, which were particularly high among the intelligentsia, and migration. The newly acquired German territories had to be resettled by Poles, including some from the lands annexed by Russia, as most Germans had fled or had been expelled. The loss of German labour, especially skilled artisans, meant that large new industrial cadres had to be trained.

Incorporation of the German part of the Upper Silesian coalfield and the deposits around Waldenburg (Walbrzych) gave Poland some of the largest coal reserves in Europe, since enhanced by major discoveries in the south-east near Lublin. Additional lignite deposits were also gained in the German territories, whereas in Lower Silesia substantial copper resources have been found. Zinc and lead also occur near the Upper Silesian coalfield, but no additions have been made to the modest reserves of iron ore. Other resources include sulphur, salt and potash. Some petroleum deposits were lost in the territories taken by the Soviet Union, though both petroleum and natural gas occur in the remaining territory, unfortunately in deposits too small to cover all domestic needs.

In German Upper Silesia, a number of important iron and steel plants were added to the capacity elsewhere in Poland (notably Częstochowa), and the large *Lenin* plant at Nowa Huta east of Kraków was built during the 1950s' search for autarky. Home coals are not outstanding coking quality and are mixed with imported coking coal, while considerable imports of iron ore are made, especially from the U.S.S.R. The Upper Silesian coalfield, with its large thermal electricity generators, has become a focus for non-ferrous electrometallurgy (e.g. aluminium, using Hungarian bauxite). The annexed German territories in Silesia also added a substantial chemicals industry, while German wartime construction had developed the Oświęcim plant. More recently, petrochemicals have begun to develop along the crude oil pipeline from the Soviet Union across Poland to the G.D.R., and the new Baltic oil terminal east of Gdańsk may also stimulate petrochemicals on the lower Vistula.

Territorial change added to the engineering industry, though towns in the prewar territory of Poland are among the largest centres. Nevertheless, the incorporation of Gdańsk and Szczecin have greatly augmented shipbuilding capacity (Poland is one of the main producers in Eastern Europe). The expansion of engineering has included diversification into such branches as motor vehicles, though whether such products are really suited to the Polish economy is questionable. The existing Polish textile industries around Bialystok and Lódź (over 40 per cent of output) were augmented by mills in former German Lower Silesia, though the industry is largely dependent on Soviet raw materials.

Despite big investments in industry, Poland has remained an economy with a strong agricultural element (30 per cent of the population is still engaged in farming). The 'classical' form of Soviet-type collectivisation was early rejected and the peasants have been left remarkably independent, with encouragement to develop co-operatives. Marked contrasts in farming exist between the former German lands and the rest of Poland: in the former state farms are most common because there have been difficulties in resettling the country-side. In general, standards of farming fall towards the east and the best land is not always optimally used. The performance of farming has been unimpressive: holdings remain small and machinery and fertilisers have been inadequately available. Farm produce has, however, been a significant element in Polish trade with the West, though the European Community Common Agricultural Policy has raised increasing barriers.

Poland occupies an important place in the transport system of Eastern Europe, especially through its plethora of ports added by the territorial changes in 1945, to which it has sought to attract traffic from the interior, including the Danube basin. The railway system provides important north–south links from these ports inland, but also east–west routes from the Soviet Union to Western Europe. The big investment in railway modernisation was not until recently matched by similar development of road haulage, as elsewhere in Eastern Europe. Poland could play a major role in the development of the Eastern European waterways system, if a heavy capacity route from the Dnepr basin to the Elbe and Danube via the Vistula basin were to be built.

The significant role played by regional planning in the socialist *bloc* countries of Eastern Europe has been reflected in Poland by attempts to disperse industry from the main agglomerations into rural areas deficient in economic opportunities. During the search for self-sufficiency, the large integrated *Lenin* works in the new town of Nowa Huta east of Kraków was built to 'soak up' rural underemployment, while prewar attempts to spread industry more evenly in southern Poland have been continued. In the annexed German lands, nearly all industrial towns have grown beyond their prewar population size, whereas rural population is still below that level over much of the territory.

Czechoslovakia

The most industrialised Slav country before 1939, Czechoslovakia did not fall fully into the Soviet orbit until the *coup d'état* of 1948, and its economic progress has been slowed by problems of modernisation and investment dilemmas in regional policy. A long sausage-like territory compared with the compact rectilinear Poland, territorial loss in 1945 comprised only the small upland tail of the Carpatho-Ukraine. Prewar like Poland, it contained large minorities, of which more than three million Germans were the biggest. The particularly sour relations between Czechs and Germans after 1938 resulted in the latter's expulsion in 1945, depriving the Czechoslovak state of a significant part of its industrial labour force and its leading entrepreneurs, while many industrial towns were left virtually deserted when the Germans had gone. Training a new labour force and redeveloping the deserted towns had similarities to the Polish problems in the annexed but depopulated German territories. The substantial share of national resources needed to achieve this redevelopment in Bohemia–Moravia aroused renewed and increased suspicion in Slovakia of the goodwill of the Prague government towards Slovakian underdevelopment, a question that had discoloured Czech and Slovak relations before 1938. Investment diverted to Slovakia, however, held back resources for modernisation, re-equipment and training in the main industrial concentrations in Bohemia–Moravia where new wealth could best be generated.

Czechoslovakia is one of the major heavy industrial producers in Eastern Europe on much the same scale as Poland. The main coal deposits lie in Czech Silesia, but others occur between Prague and Plzeň as well as in northern Bohemia. Lignite occurs in the north-west around Most and in Moravia and Slovakia, and there is some petroleum on the Austrian border in the Moravia basin. Modest iron ore deposits occur in central Bohemia and in eastern Slovakia, but most ore has to be imported. Copper, lead, zinc, silver and other metals occur in numerous but small deposits. The iron and steel industry is concentrated in Czech Silesia and around Plzeň-Kladno in the west. Early postwar Russian plans were rumoured to have considered forming Polish and Czech Silesia into a politically autonomous heavy industrial core for Eastern Europe. Older and smaller plants in Slovakia have been overshadowed since the late 1960s by the large new works at Košice, supplied by a special broad gauge railway with ore and coal directly from the Soviet Ukraine.

As arsenal of the Habsburg Empire, Czechoslovakia had inherited a substantial engineering industry (including the large *Škoda* works at Plzeň), and in the 1920s developed its own indigenous motor car industry. With a significant heavy engineering sector, it has become an important adjunct to the economic strength of the Soviet *bloc*. There is also an important chemicals industry concentrated along with engineering in Bohemia–Moravia (the Czech lands). The energy problem has been eased by the building of natural gas and petroleum pipelines from the Soviet Union, which has stimulated development of petrochemicals. The emphasis on capital equipment in the economics of Eastern Europe since 1945 and the low priority accorded the individual consumer has had an adverse effect in Czechoslovakia. The considerable glass industry of northern Bohemia suffered through its low priority and the loss of its labour force with the expulsion of the Germans (who now compete from new works in West Germany). A low priority was given to the Bohemian textile industry, because of its consumer orientation and because textiles have been among the more appropriate industries to develop in Slovakia as part of the equalisation of regional development. Also encouraged in Slovakia has been light engineering assembly work (notably in the Váh valley). Nevertheless, Czechoslovak consumer industries – textiles, clothing, shoes and domestic equipment – have become highly significant in *Comecon*, though their growth might well have been larger in a free market economy or under other priorities.

In a central position in Europe, Czechoslovakia is an important transit land – the 'turntable of *Comecon*' – for railway traffic, though most domestic freight traffic is on an east–west axis between the main industrial concentrations. The country has also falteringly taken up completion of the long-standing project of the Oder–Danube Canal through the Moravian Corridor, while a branch to the Elbe is also planned. From its inland position, Czechoslovakia is forced to use other people's ports – much traffic goes to Polish and East German ports, some to Hamburg in West Germany, and Yugoslavia has sought to attract traffic to Rijeka. Road traffic has grown owing to development of an extensive bus system, while a motorway from the East German frontier via Prague to Bratislava is being built.

Agriculture plays a lesser role in employment in Czechoslovakia than in most other Eastern European countries, but it is still more important than in Western Europe. In Bohemia–Moravia, farming is often part-time, with industrial employment as the mainstay, whereas economic policy has sought to reduce the dependence of employment on agriculture in Slovakia. Czechoslovakia, unlike Poland, has remained close to Soviet practice, and collectivisation has brought change to the landscape and settlement pattern. Unfortunately, with so much upland and mountain farming, returns over large areas are mediocre. Within *Comecon*, beet sugar and hops are important Czechoslovak contributions, while the large forests supply a significant paper-making industry.

German Democratic Republic

Usually considered now to be the tenth major industrial country in the world, the German Democratic Republic has become one of the key members of *Comecon*. It is without the strong heavy industrial emphasis of Poland and Czechoslovakia, but because of the size and advanced technology of its chemicals industry in the Elbe–Saale basin and its sophisticated engineering industries, particularly in the Saxon lands and Berlin (electrical engineering), it is a major contributor to the overall *Comecon* economy. The G.D.R. optical goods industry of Jena and Dresden is the chief centre for such high quality products in the whole *bloc*. As part of occupied Germany in the Soviet dismemberment of the *Reich*, the G.D.R. suffered heavily from reparations dismantling, while its whole spatial structure was disrupted by the new border on the west and by the westward shift of the Polish frontier. Isolation from the West also dislocated long-standing economic linkages. In the early postwar years its economic development was therefore retarded, but from the late 1950s it became another if more modest 'German miracle' and has overtaken rival Czechoslovakia in several sectors.

Little bituminous coal is available and the main energy source is electricity generated by power stations fired from large deposits of lignite. With limited oil and gas deposits, supplies have to be overwhelmingly imported, a task eased by completion of the Soviet pipeline to the oil refinery at Schwedt, whereas a marine oil terminal is in use at Rostock. Nuclear raw materials in the Gera and Aue districts have so far been mostly exported to the U.S.S.R., but plans exist for nuclear power stations. Iron ore deposits are poor, but there is some nickel and zinc and significant resources of copper (Mansfeld). There is a particularly generous endowment, however, of chemical salts.

The lack of coking coal (attempts to coke lignite failed) and iron ore made Stalinist policy of self-sufficiency by development of iron and steel making a fiasco. Of two large works, Eisenhüttenstadt and Calbe, built under this policy, the latter has already been closed. Specialised rolling and quality steel production has been more successful. Engineering capacity has been geared to the heavier types of equipment required in *Comecon*, but several branches were disrupted until that part of their supply of components remaining in West Germany could be replaced. Like all G.D.R. sectors, labour shortage arising from the strong migration of people to West Germany before 1961 (the Berlin Wall) has affected engineering assembly industries badly, leading to co-operation with Hungary, where rural labour is plentiful, to carry out assembly of German-made parts. Soviet emphasis after 1959 on development of the chemicals industry favoured the large potential capacity of the G.D.R., where five plants in Halle, Bitterfeld, Wolfen, Merseburg and Piesteritz supply 40 per cent of output. Synthetic rubber, artificial fibres, pharmaceuticals, and photographic chemicals and materials are major products. Originally using largely lignite synthesis, there has been a shift to petrochemicals, but a new problem is shortage of process water. With electricity comprising over one-third of costs in many plants, the industry tends to cling to sites near large generators.

The G.D.R. inherited a large consumer goods industry, including fine textiles in Saxony, but further development has received a low priority and there has been a shift to medium-quality goods to suit *Comecon* markets. Nevertheless, consumer goods – including domestic electrical equipment – remain important for the *Comecon* market and quantities have been sold to West German mail order firms.

Prewar farming standards in the G.D.R. were well in advance of most of the remainder of the present *Comecon* members, but land reform and collectivisation in the immediate postwar years caused disruption, so that well into the 1950s yield and productivity lay behind even mediocre prewar years, creating a serious food supply problem. Following Soviet principles of collectivisation closely, farming is now among the more efficient systems in *Comecon*, though the proportion of total employment is lower in the G.D.R. Whereas farm produce forms an important export item in most Eastern European countries, the G.D.R. needs to augment home production by imports.

Despite its central position, the transport system of the G.D.R. was badly dislocated by the dismemberment of the *Reich* and by reparations dismantling. The reconstruction since the mid-1950s has sought to adjust to the new spatial pattern, with freight concentrated in the Elbe–Saale basin and Saxony and on lines to adjacent *Comecon* members. The G.D.R. has also tried to attract traffic from Danubia and Czechoslovakia to Rostock in competition with the Polish ports. The Elbe waterway and the Mittelland Canal have been reduced in significance as traffic arteries, owing to their being cut by the frontier with West Germany. The G.D.R. has the densest and best road system in Eastern Europe, including several prewar motorways which, in the early 1970s, were extended and missing links added, though road haulage remains relatively limited. The most important transport project has been the plan to build a high capacity railway, a motorway and possibly a canal from the Berlin area to Rostock port. Disruption of the transport network for all media because of the isolation of West Berlin has meant costly construction of diversionary railways, roads and even a canal.

After a massive increase in the G.D.R. population through refugees and expellees in the immediate postwar years, the sovietisation of the economy and society generated an intense outflow of people to West Germany. As these people were principally in the demographically and economically most productive age groups, the impact generated an incipient declining and ageing population, with growing labour shortages and serious age–sex imbalances. Migration of non-German labour into its territory, like the 'guest workers' of West Germany, has been unacceptable in the Marxist–Leninist setting, impeding economic development. It has also coloured the relatively low level of change between employment proportions in agriculture and industry, again unlike the rest of Eastern Europe.

Although Marxist–Leninist planning dogmata stress the importance of the town, a high level of urbanisation existed even prewar in the G.D.R.: for this reason, the considerable rise in the proportion of urban population typical of postwar Eastern Europe has not been marked. Postwar policy has been to switch development into medium and small towns, rather than accord high priority and deflect valuable resources to rebuilding badly damaged major towns. This has also helped to equalise living standards between town and country by stressing development in many small northern towns, formerly largely agricultural markets. New towns have also been built to serve new industrial sites, like the developing Lusatian brown coalfield, the Oder valley (for example, Eisenhüttenstadt, Schwedt) or for new expansion in the Elbe–Saale basin (for example, Halle-Neustadt for chemical workers).

Hungary

Much of Hungarian industry had lain in territories lost in 1919 and the country, saddled with an unbalanced economy, was unable to make much progress in the troubled interwar years. Economic policy since 1945 has sought to rectify this imbalance and to expand industry to absorb rural overpopulation, though the spatial pattern remains unduly dominated by Budapest and Hungarian agriculture still employs about 30 per cent of the labour force. The major role was played prewar by vast estates that have now been turned into state farms and collectives, though these enjoy considerable flexibility. Every effort has been made to develop commercial production, shifting to livestock, fruit and special crops that find a substantial export market. This policy has succeeded because Hungary's crops mature after the Southern European earlies and before the Central and North-west European main crop. Nevertheless, like Poland, Hungarian exports have suffered from tariff barriers erected by the European Community Common Agricultural Policy.

While there is plenty of brown coal, bituminous coal is scarce, though some is suitable for metallurgical use. Home supplies of natural gas are augmented by supplies from Rumania, while home oil resources have been developed in the south-west, and crude oil is now drawn by pipeline from the Soviet Union. Plans exist to develop the hydroelectric potential of the Danube and Tisza. Some iron, manganese and copper ores are worked, but the main resource is bauxite, whose processing raises problems because of its energy-intensive nature. Increasing quantities of the ore are processed at home, but various arrangements for processing in Poland, the Soviet Union, the G.D.R. and Czechoslovakia, all more fortunately placed energywise, exist.

Cut from their ore supplies in Slovakia in 1919, several small Hungarian ironworks between the wars had stagnated. Under the autarky policies of the Stalinist period a large new plant, using Soviet, Bulgarian and Hungarian raw materials, was built on the Danube at Dunaújváros. The main emphasis has been on engineering, especially transport equipment and electrical goods, mostly concentrated in Budapest, while the textile industry has been re-equipped. An unusual but useful contribution is made in pharmaceuticals from vegetable bases.

Regional policy has concentrated on reducing the dominance of Budapest, with industry decentralised into country towns where labour is plentiful, but a belt from Miskolc via Budapest and the bauxite fields of the Bakony to the south-western oilfields around Nagykanizsa still dominates the picture. If traffic were to develop more strongly on the Danube (for example, after completion of the Main–Danube Canal), Hungary would occupy a commanding place on the river, though it already provides important railway transit. Hungary's development is, however, constrained by the tensions of its long frontier with the dissident Yugoslavia and the West-oriented Austria, though it has always tended to 'think German' in its contacts.

Rumania

Though in many respects still underdeveloped and backward, Rumania has possibly the greatest long-term potential of all South-east European countries. Using the sound endowment of mineral wealth, the government has sought to maximise industrial growth at the expense of living standards, without the turbulent repercussions experienced in the more advanced economies. Rumanian industrial growth rates have not only been among the highest in *Comecon* but also among the highest in the world. Rumania has pursued industrialisation ruthlessly and defiantly against attempts by *Comecon* to turn it into a supplier of raw and semi-processed materials to the more industrial members.

Before 1939 Rumania was the main European producer of crude oil, but its relative position declined as output failed to grow in step with world trends. The main fields remain along the Carpathian foot near Ploieşti, but considerable resources of natural gas have been developed in Transylvania. Deposits of hard coals, reasonable in quantity and quality, occur in south-west Transylvania, whereas the main lignite fields lie north-east of Oradea. Efforts to develop hydroelectricity in the Carpathians and in a big project with Yugoslavia at the Iron Gates of the Danube have been reasonably rewarding. Rumania has one of the least critical energy supply problems in Eastern Europe. A wide range of metallic ores occurs: iron ore is mostly in the south-west, though imports have also to be made; manganese occurs in the north: there is also some chrome. A little bauxite is found, but most important are polymetallic ores (including copper, lead and zinc), while it is reasonably well endowed with silver and gold, of which it is Europe's main producer. The Bihor Massif is most important for non-ferrous metals. There are also large resources of salt and pyrites.

The iron and steel industry has been greatly expanded (though half the ore requirements are imported) in the Reşita–Hunedoara area, and several other centres roll steel. Plans to build a large integrated works using mostly imported raw materials at Galaţi have been attacked in *Comecon*, where Poland and Czechoslovakia would like any further expansion in capacity to take place in or near their own main producing districts. There has also been expansion of non-ferrous metal processing: Baia Mare is the main centre. Aluminium is produced at Slatina in the Argeş region.

Plentiful raw materials have been an important base for development of the chemical industry, with an emphasis on artificial fertilisers, but also developed have been synthetic fibres and pharmaceuticals, as well as heavy industrial chemicals. Emphasis has been laid on petrochemicals in the Ploieşti region. Long term, Rumania could become a major supplier of industrial chemicals to the socialist *bloc*. Rumania already has the largest production of electrodes, sharpening stones and other abrasives in South-east Europe, while it is a major producer of carbon black.

Development of an engineering industry has been an important adjunct, but in such sectors as vehicle building it is questionable if the right economies of scale can be achieved to avoid waste. Rumania specialises in petroleum working equipment, whereas it has made use of licences from Western producers to build diesel engines and similar equipment, and it is the largest producer of railway wagons in Eastern Europe alongside Poland. Much emphasis has been laid on farm machinery construction. Engineering is notably concentrated in the largest towns and it has been another sector where development has been more vigorous than *Comecon* plans have foreseen. Rumanian petroleum engineers and technologists have helped to develop oil production in other *Comecon* countries – such as the G.D.R. and Bulgaria– and in the Third World (for example in India and Afghanistan).

Despite the massive industrialisation, agriculture still employs about half the population. Collectivisation was applied slowly, so as not to disrupt production, which has risen modestly. Although climatic and soil factors over large areas favour farming, historical factors combined to make it backward, as well as generating a society that hindered peasants improving themselves. Rural overpopulation, poverty and malnutrition were major problems to overcome, whereas other special difficulties were created by the large areas of upland pastoralism. Despite such failings, Rumanian agriculture has been able to produce substantial surpluses and even to 'loan' grain to the Soviet Union when its crop failed. Nevertheless, though foodstuffs comprise about half the country's exports, there is still a need to increase output by shifting from extensive methods (including grain culture) to more intensive forms of farming, for which the country is admirably suited.

The change from an overwhelming agrarian economy to a more balanced situation between agriculture and industry has meant much change in spatial pattern of the country, but this has been achieved by spreading development fairly widely and has been without a major drift from the countryside to the few main industrial towns. Consequently small country towns have grown more than the great towns, although over half of all industrial output still comes from five main districts – Ploieşti, Hunedoara, Banat, Braşov and Bucharest. The policy of spreading development has demanded heavy infrastructural investment in such things as water supply, electricity distribution and transport in predominantly rural areas before the new industries become productively effective. Equally costly has been establishment of adequate training and educational facilities. Transport investment has aimed at increasing the capacity of the railway system to carry rising industrial freights and at developing the network to help weld the country together across the obstacle of the Eastern Carpathians. Road transport has played a minor role, though attention has been given to improving Danube navigation and increasing the harbour capacity of Constanţa, Braila and Galaţi.

Bulgaria

Bulgaria long remained one of the less developed parts of Europe. Coal and lignite occur in the west and centre, while small deposits of oil and gas have been found in the north. Iron ore, manganese and chromite occur, as well as molybdenum, copper, lead and zinc – all metals much in demand in *Comecon*. Investment has also been made in hydroelectric generators. Since 1945 industrialisation, given a high priority, has been impeded by a lack of skilled workers and has depended on help in training from other *Comecon* members.

Industrialisation begun under the Stalinist autarky policy concentrated on the heavier branches. Using mostly home raw materials but some imported coking coal, an ambitious plan for iron and steelworks built plants at Pernik and Kremikovtsi. Of almost equal importance is the preparation of non-ferrous metals, notably electrolytic copper, lead and zinc. Sheet copper is an important export. Engineering now employs as many people as all pre-1939 Bulgarian industry, though well over one-third of the output comes from Sofia alone. Transport equipment and farm machinery are the main items: forklift trucks are exported to Britain. With considerable help from the G.D.R., local raw materials have been used to develop heavy chemicals, especially artificial fertilisers and agricultural chemicals. Petrochemicals have been developed at the Burgas refinery, using Soviet and Algerian crude oil, whereas Jambol has a synthetic yarn factory. Traditional handicraft industries and related factory industries such as textiles remain significant.

Over one-third of all Bulgarian exports are derived from agriculture and almost one-third of total industrial output is from food processing plants. Unlike some *Comecon* countries, where postwar land reform broke up big estates, Bulgarian land reform reduced the vast number of small land parcels and tiny farms into more useful moderately sized collectives, whose inception was followed by extensive ameliorative work, including irrigation works associated with large hydroelectric barrages in the mountains. Summer aridity is a major hindrance to farming. Although retaining conventional collectives (in recent years closely related to processing plants), the record of productivity has been better than in any other Eastern European country. Bulgaria is fortunate, however, to be able to rear crops impossible to grow successfully elsewhere in Eastern Europe. The country could become one of the major sources for the rest of the *bloc* of many vegetables, fruits, tobacco and such special crops as attar of roses.

Albania

In Albania, the most backward and underdeveloped part of Europe in 1945, Communist control set under way a policy of modernisation and development based on work already started before the war by the Italians. The main contribution of this small country to *Comecon* was its mineral wealth – chrome and iron–nickel ores, copper and asphalt. Oil, natural gas and some brown coal form its energy base. With the help of more developed countries, notably Czechoslovakia, this wealth was opened up during the 1950s. Simple engineering, textile manufacture and food processing have been started, but during the autarky phase plans were generated to develop iron and steelmaking. All development has been hampered by poor transport and, in relation to its size, it has had one of the most rapidly growing railway systems in the world, but there has also been extensive roadbuilding. Plans exist to link its railways to the new Yugoslav Belgrade–Bar railway. Distinguished by its extremely high rate of population growth, it is questionable whether economic development can keep pace with this demographic explosion. The shift of Albania away from Soviet influence into the Chinese camp since the early 1960s has lost for the Soviet Union a useful position against the dissident Yugoslavia and a command of the entrance to the Adriatic.

Yugoslavia

Since its clash over political issues with the Soviet Union in 1948, Yugoslavia has been a dissident in the Communist camp. With less than half its boundary in the Soviet *bloc*, it was in a stronger political–geographical position to dispute issues with its Soviet mentor than any other country. Its ambivalent political relationships between East and West have greatly influenced the development of its internal spatial pattern. Although not a member of *Comecon*, Yugoslav policy has followed similar objectives: rapid industrialisation, urbanisation and the equalisation of regional development.

Of all Eastern European countries, Yugoslavia is one of the best endowed with minerals and is a major European producer of non-ferrous metals. Iron ore, chrome, manganese and tungsten occur, while lead, zinc and mercury are significant and important copper deposits are worked. There is a considerable potential for bauxite working. Some gold, silver and uranium occur. The energy resources are reasonable – coal, lignite, natural gas and oil. The natural resources form a promising basis for industrialisation, though for iron and steelmaking good-quality coking coal is imported. The emphasis on industrialisation has been in the heavier branches to use these minerals. The older small ironworks in the north of the country have been augmented by newer and larger plants, notably Zenica in Bosnia and Skopje in Macedonia. There has also been modernisation and development of non-ferrous metal processing. Engineering developed mostly in the more industrialised northern towns and in Belgrade, whereas on the coast – notably in Rijeka and former Italian ports – shipbuilding has flourished. Motor cycles and bicycles are exported, while motor vehicles are also assembled – the main plant licensed by *Fiat* at Kragujevac in Serbia. A scatter of electrical engineering plants has developed, with goods assembled under licence from Western European firms, and radios and televisions made in Niš. Home raw materials provide the basis for the chemicals industry, though this has not been affected by the hothouse atmosphere of development pursued in *Comecon* under Soviet pressure. The plants are mostly in the north and produce industrial chemicals like soda products, sulphuric acid, but also agricultural fertilisers. With extensive limestone areas, Yugoslavia has developed a large cement industry and is a major exporter to the Mediterranean basin. Its considerable handicraft and consumer goods industries have found useful markets in Western Europe (for example, leatherware).

Because of the large extent of poor land, agriculture presents some of the greatest economic problems in Yugoslavia, particularly as it remains a major source of employment, especially in the poorest districts. The main arable areas are in the Sava–Drava lowlands and the Danube–Tisza mesopotamia. Pastoralism with patches of arable land predominate in the uplands and mountains, where there has been severe overgrazing. Collectivisation, a basis of disagreement with the Soviet Union, was rejected more on practical than philosophical grounds and peasants left to farm privately, though encouraged to form co-operatives. The remaining socialist sector of agriculture has made the best showing, largely because it holds the best land (vacated by the fleeing German minority in 1945) in the Vojvodina. The traditional pig rearing of the grain lands and more productive uplands has been linked to a modern meat-processing industry that exports considerable amounts, mostly to Western Europe.

Like other Balkan countries, Yugoslavia has put emphasis on completing and modernising its railway system: one of the largest postwar European railway projects has been completion of the Belgrade–Bar railway, giving Serbia a direct outlet to the Adriatic. Unlike *Comecon* members, Yugoslavia has given higher priority to road construction to help tap the lucrative tourist traffic from Western Europe to the Adriatic coast and Greece. It has also tried to develop its ports on the Adriatic to serve the Danubian interior in competition with East German and Polish ports, but a serious constraint has been the poor transport links across the Dinaric mountains.

The disparity of economic development between north and south in Yugoslavia is substantially greater than the regional contrasts in Bulgaria or Rumania. Underdeveloped areas, covering 40 per cent of the national area and encompassing 30 per cent of the population, extend across the centre and south of the country. These are areas with little employment other than in farming and high rates of natural increase. Heavy investment to develop these regions deflects scarce resources away from the more favoured areas which supply many of the goods needed for development in the less favoured areas: growth in the more favoured areas is therefore slowed and inflationary pressures besides political strains build up.

The policies of the European Community and *Comecon* applied for nearly a quarter of a century have come to have an impress upon the spatial pattern and landscape of Europe. The European Community has sought supranational integration, with a new attraction being given to frontier areas between states as mobility was freed: in contrast, *Comecon* has made an international approach, ostensibly protecting national sovereignty, with protection of national spatial structures and a low mobility of finance, resources and labour. Nevertheless, new spatial structures have begun to appear, owing to industrialisation and urbanisation in Eastern Europe and the policy of attempting to equate intranational regional development. Change may accelerate as a new supranational emphasis began in *Comecon* in the 1970s, which seeks to speed up establishment of large-scale sophisticated plants to achieve optimum output. A 'supranational' element has been present since the inception of *Comecon*, owing to the disparity in resources and influence between the Soviet Union and its 'partners', whereby Soviet needs have strongly conditioned the patterns of trade and even economic development in Eastern European members. Soviet influence has been strongly pitched against regional groupings in Eastern Europe, so that concepts such as Danubia or the Balkans in economic and political–geographical terms have now relatively little significance.

References

1. An excellent small atlas of the Community is *The European Community in Maps* (Ed. I. B. F. Kormoss), Press and Information Service, European Communities, Brussels (1974); also *Atlas of Europe – A Profile of Western Europe* (Ed. G. S. Browne), Bartholomew and Warne, London (1974).
2. A small atlas of the Eastern European *Comecon* members and other socialist states is *Thematische Karten zur Welt von Heute – Teil VI: Sozialistische Staaten Europas (ohne UdSSR)*, Verlag Bagel, Düsseldorf (1972).

8 Towards the Year 2000

The Second World War was followed by a polarisation of the world around new hearths of economic and political power in the United States and the Soviet Union, as the already shattered European Age finally crumbled and Europe itself was torn in two as the two new and apparently incompatible political spheres confronted each other. From this confrontation new groupings of Europe's nation-states have arisen in the two major economic and social communities outlined in chapter 7, and through their influence Europe will look quite different at the threshold of the twenty-first century compared to mid-twentieth century. In the changing world, Western Europe, alongside the United States, no longer constitutes the sole foci of advanced industrial economy and society – Japan and the Soviet Union are rapidly becoming serious contenders, and this type of economy and society will have spread to a much greater part of Europe by the year 2000 compared with the interwar 1930s and even the 1950s and 1960s. By setting the trends of these periods and those just beginning to emerge towards the 1980s in a geographical matrix, it is possible to try to suggest some broad characteristics of Europe 2000. With so many variables and so many imponderables, there are, no doubt, alternative interpretations at many points; but without being too dogmatic, it is hoped the general outlines of the prediction will not fall too far from the reality of the year 2000 that is already beginning to assume some mystic aura.[1]

Interwar observers had seen Europe divided into two main economic spheres – one predominantly industrial, mostly in Western and Central Europe, but the other an areally larger sphere characterised by an economy dominated by peasant farming that typified Eastern and Southern Europe. By the 1970s, this distinction, though never crystal-clear in definition, had become blurred beyond recognition. Despite continued decline in the employment share of farming in 'industrial Europe', its distinction was being eroded by the rapidity of industrialisation in areas previously perceived as belonging without doubt to 'agricultural Europe', notably the socialist *bloc* countries of Eastern and South-eastern Europe. If the trends of the 1970s continue to the year 2000, there will remain little difference between most European states in the basic role of farming and industry in their economies. This does not suggest that the economies of Europe's nation-states will be the same – industrialisation in the socialist *bloc* countries, based on Soviet experience, will clearly be different to the industrial patterns in the Western free market economies, though the present distinctions may become less over time. Indeed, policies in both East and West tend to stress the equalisation of economic levels at regional level and between nations. A further element of change has been the growth of the tertiary (and even quaternary) economic sectors in the most advanced countries: if this trend is pursued, a grouping of states – probably Germany, France, the United Kingdom and Benelux – may by the next century form a new truly 'post-industrial Europe',[2] offering indispensable and sophisticated services to the remainder.

The ideological split of the Cold War drew Northern and Western Europe (and even part of Southern Europe) into a close military and political association within the North Atlantic basin. Economic rehabilitation and growth, as much influenced by technological change as by political commitment, developed within the same spatial pattern. On the other hand, Eastern and South-eastern Europe (though there are some loose stones in the structure) have been encapsulated by the economic and political system with continental dimensions erected in Eurasia by the Soviet Union. As American and Soviet interest in Europe wavered with the emergence of new problems elsewhere in the world in the third quarter of the century, there has been a growing contact and a search for *détente* between these two spheres across the ideological divide. This desire for contact doubtless arises from the long-standing appreciation in Europe that all its parts are mutually interdependent – as the status and influence of the European nation-states within their respective *blocs* have grown, they have brought pressures to bear to re-establish closer trans-European ties (haste in this process, however, caused such an aggravation in the Soviet Union that it triggered the 1968 Dubček crisis in Czechoslovakia). It seems reasonable to assume that this process has now gone so far as to be virtually irreversible and that technology and economics make armed conflict ever less attractive, so that there will be a continuous if erratic movement towards closer and more intricate intra-European ties throughout the declining years of the present century, provided a reasonable economic and military balance is maintained. As *détente* in every sense becomes more real, Europe is in a commanding position, straddling two different economic and political systems, able to benefit from the extensive spectrum of interchange possible along the interface between them. Do we see in this situation the possible role of Europe as the Third Force in the global political–geographical structure, as envisaged by many supporters of the European Movement in the late 1940s?

The economic sphere of maritime Europe, focused upon the North Atlantic basin, is in effect a strengthening of a long-standing trading sphere by new commercial links in a political and defensive setting. Europe accounts for a significant part of the formidable array of workshops set around the basin, for in Western Europe and North America accessible to the Atlantic are some of the largest concentrations of industry in the world. The growth from the late 1950s of bulk oceanic carriers of low unit freight cost have greatly enhanced the position of the coastal and deep-water heavy industrial element in these agglomerations. Nevertheless, despite the advantages of such bulk carriage, by the early part of next century there may be a tendency for parts of this sphere to depend to a rising degree on basic heavy industries using the raw materials base of interior Eurasia, if availability of minerals in cost terms comes to favour the latter. The workshops of the Atlantic basin will no doubt show a continued shift towards high technology, in which the bonds established across the ocean by immense multinational companies will be significant (though possibly only with government assistance) in supporting the expensive research and development programmes demanded. Although increasing activity to develop multinational concerns has taken place in Europe – aided by legislation of the Economic Community – experience suggests that North America will remain the vanguard of this trend. Extrapolating from the present situation, the Atlantic peripheries could become the focus of the 'post-industrial' economies, dependent to a substantial degree on tertiary (and perhaps even quaternary) activities, with obvious scope for linking the technological skill of the European littoral workshops with the labour reserves of the interior, particularly in Eastern and Southeastern Europe.

A combination of political factors reducing the ideological and consequent physical isolation of the Soviet interior of Eurasia and trends in modern technology have begun to open a promising landbridge between the major Atlantic economic sphere and the rapidly growing Pacific Ocean economy, incorporating the great potential of South-east Asia. Industrial production in the continental Eurasian sphere shows a strong concentration in the western part – the workshops of Eastern Europe and the western Soviet Union account for four-fifths of the total socialist *bloc* output. Indications from the late 1960s onwards suggest that this western concentration is becoming more significant and has deflected Soviet investment into long-neglected western frontier areas, like western Byelorussia and the western Ukraine, which might otherwise have gone into Siberia or Central Asia. An especially meaningful grouping in Eastern Europe, in many ways very similar to the *Montandreieck* of the European Community, is formed by the heavy chemicals industries of the Elbe–Saale basin (G.D.R.), the varied heavy industries of Bohemia–Moravia and an expanded Upper Silesian industrial district in Poland, all well placed for closer co-operation with Western Europe, and in several respects their structures supplement those of Western Europe. The third *Montandreieck* is essentially Soviet – the Industrial South (that is, the Ukraine), the Industrial Centre (Moscow and environs) and the Ural–Volga region – but has growing contacts with the two already defined.

The great strength, in the long term, of this continental sphere is that *Comecon* encompasses vast potential mineral wealth in Siberia and Soviet Central Asia. Although the Soviet Union has told the Eastern European members that they cannot expect to remain favoured indefinitely, they will nevertheless remain advantageously placed in a situation of rising shortages. Siberia has no doubt a most significant role to play in Europe – both East and West – as a source of energy, whether for oil or gas, or also electricity from water, thermal or nuclear generators. Energy supplies will become increasingly critical in Europe around the end of the present century and local resources appear puny compared with the Siberian potential. The large diameter pipeline or the super-high tension transmission line could play as vital a role in European energy supply in the year 2000 and beyond as the mammoth tanker does in the 1970s. The development of these Eurasian energy and raw material supplies will demand financial and technical wealth: the Soviet Union already feels itself taxed to undertake the task alone and the Eastern European members of

Comecon have not the financial or material resources to contribute much, so that it may rest upon aid from advanced Western economies with wealth and know-how, as already initiated by the United States, France and Japan. West Germany has also explored possibilities of obtaining energy from this sphere. It may possibly be a way to ease the problems of Western Europe, Japan, and even the United States faced by constantly rising OPEC oil prices.

Eastern European members of *Comecon*, by early next century, may well be significant suppliers of manufactured goods to Western European markets – several indicators point to a possible wide range of domestic goods, consumer durables and even industrial components that could possibly be more advantageously produced in Eastern Europe than in Iberia and Southern Europe. The more sophisticated goods could be supplied from the German Democratic Republic, Poland and Hungary, or Czechoslovakia, whereas simpler products might be supplied from South-eastern Europe (for example, textiles, footwear, etc.). On the other hand, a wide range of articles made in Western Europe, including many consumer goods as much as the products of high technology, has an immense potential market inside *Comecon*. The growth of this pattern of trade had already begun by the end of the 1960s and, without political constraints, could be expected to develop vigorously. Historical evidence suggests that the reintegration of overall European trading patterns is carried through as soon as conditions allow after disruption or isolation.

Regional Realignments for the Year 2000

The political–geographical shifts in the twentieth century have created a number of new regional concepts and relations, though perhaps less radical than might have been expected from the great upheavals caused by the world wars. The principal question is whether further changes in existing functional regions may be expected as the new century is approached, especially through the impact of a widening inter- and supranationalism.

Northern Europe – Scandinavia – has been little changed, though Finland lost territory to Russia, with which Norway now has a common frontier in the far north. Economically Scandinavia remains dominated by Sweden, one of the world's apparently most prosperous and successful countries, but the long-standing neutralist policies of Scandinavia, originally breached by Norway and Denmark joining NATO, have been further holed by Denmark formally joining the European Community, acknowledging the reality of its economic ties. In contrast, Norway rejected formal membership, though development of its North Sea oil resources may well draw it increasingly towards the Community. Heavily dependent on fishing in an age of depletion of stocks and pollution, Iceland is an Achilles' heel in the overall Scandinavian economy, but its apparent conflict with some Western European states over the fishing issue may be a reflection of a new identity arising between the maritime states, inside and outside Scandinavia, over the need to protect their fishing interests by fresh interpretations of sovereignty over the sea in much expanded territorial waters.

The easing of tensions in the late 1960s helped to recreate the life of the Baltic, though even during the Cold War period a considerable volume of commercial circulation continued on this sea, with a large potential for interchange between Scandinavian high technology industries and the energy and raw material supplies available in *Comecon*: the close relation of the Finnish and Soviet economies, generated first in a draconic reparations agreement, illustrates the possibilities. The quickening development of landbridge routes across Eurasia, linking 'maritime' and 'continental' spheres, puts a new emphasis on the Baltic – already trans-Siberian rail-borne containers from Japan are being shipped to European coastal destinations via Soviet Baltic ports. A major impedance to Baltic development may be the limits to physical access of large bulk carriers, though relaxed east–west relations could generate the necessary joint action to improve fairways or enlarge canal access, particularly if planned oil terminal facilities (for example, at Ventspils) generate suitable levels of traffic.

After Mitteleuropa

By the year 2000 the greatest realignments are perhaps to be expected in the main continental peninsula, already the focus of major changes after both global wars of the twentieth century. With the emergence of a strong German Empire and the powerful sphere of influence of the germanised Habsburg Empire, Europeans came to accept the concept of a well-recognised but vaguely defined *Mitteleuropa*, a massive political and economic regional 'force field' extending its influence south-eastwards beyond the German sphere into Danubia and the Balkans, just as it also penetrated into the eastern and north-eastern marchlands. Between the world wars it was much weakened, divided between a troubled Germany and the querulous Succession states, though the Nazis revived it with brief success culminating in 1942 in the declaration of Greater Germany. The collapse of Germany in 1945 left this important sphere as a vacuum to become the political and economic interface between rival ideologies during the Cold War.

The disappearance of *Mitteleuropa* brought an extensive re-ordering of the spatial pattern in Europe, undertaken with considerable vigour in both Eastern and Western *blocs*. Of the new functional regions, one of the most impressive is the Greater Rhineland, the product of the shift from the economic nationalism of the 1930s to the supranationalism of the European Community. The key to this new region was the development of the *Montandreieck* in the early 1950s through the linkage of German, French and Benelux heavy industry, further strengthened by the transport relationships generated by the creation of the enlarged Rotterdam-Europoort and the general expansion of trade and transport in the *Eurodelta*. The arterial role played by the Rhine itself within this region and within the broader framework of Western Europe has made the river the axial focus, though the regional limits have spread beyond the river's natural physiographic catchment. The Greater Rhineland has thus come to encompass several of the main industrial nodes of Western Europe – the *Eurodelta* node, the Rhine–Ruhr and Rhine-Main industrial nodes, the Neckar basin and *Saarlorlux*, besides the Swiss Mittelland. In close association though not dominantly so, Northern France, the Belgian Sambre–Meuse node, the Rhône Corridor,

Greater Paris, North-west and South Germany, form an outer ring of related industrial agglomerations.

The Greater Rhineland

Released from the 'drag' of the vanished *Mitteleuropa*, the first impetus to formation of the Greater Rhineland came through heavy industry, particularly associated with the large coal resources, though its locational attractiveness within the Economic Community has drawn in from the mid-1950s onwards an increasing variety of other industries. The energy revolution created by the cheap petroleum supplies of the 1960s has been strikingly reflected in the large heavy chemicals concentration around the oil terminals of Rotterdam-Europoort, as well as refinery development near or in the older heavy chemicals areas along the middle and upper reaches (for example, the lower Rhine, lower Main and the Karlsruhe area). As inland heavy industry has lost locational advantage to coastal sites using bulk oceanic haulage of raw materials, the inland attractiveness of the Greater Rhineland to manufacturing (notably consumer durables) and to high technology industries has grown. A major attraction has been proximity to the largest and wealthiest population agglomeration in Europe, containing over 75 million people and still expanding.

Accession of the United Kingdom to the original Community of the Six may be seen as adding additional extent to the Greater Rhineland, whose industries and commercial activities have been brought into even closer contact with similar operations in South-east England, the English Midlands, Humberside and North-west England. London, as a world banking and financial centre, further strengthens the role of the Economic Community in these fields, leaving only Zurich effectively outside. As full integration of the British economy into the wider supranational structure of the Community will need at least a decade to complete, the real impact of change in and beyond the Greater Rhineland will not be seen in true detail until towards the end of the present century, when it is likely that quite startling changes in the geographical patterns of population and economic geography, as a consequence, will have begun to emerge around the southern North Sea and the Channel, with important repercussions for the geography of the remainder of the United Kingdom (if such a title will then still be relevant!).

Successful growth and the ability to draw in more people and investment demonstrated by the Greater Rhineland poses an inherent danger for the future in accelerating and defining more clearly the 'Second Europe' – the less successful parts of a number of countries which fail to maintain levels of wealth creation comparable to the ebullient core. The social and economic problems of a Second Europe would create serious strains in the maintenance of aims of unity and integration central to the European Community. Nevertheless, the problem has already been identified and improvement of planning philosophies and procedures should, however, provide mechanisms that will make possible a redistribution of wealth and resources to even out inequalities (if not to eliminate them) in regional development levels. The idea of accepting, conceptually, a Second Europe was firmly rejected in the mid-1970s when reputedly suggested in the much debated 1976 Tindemans Report on European Union.

The North Sea Space

Significant finds of oil in the difficult northern basin have stimulated rapid economic development and in the final quarter of the century a new functional region – perhaps best named the North Sea Space – began to emerge, with impressive developments on the east coast of Scotland and in southwestern Norway, besides the Orkney and Shetland Islands, though the spin-off has also been felt in North-east and Eastern England, in Denmark, Northwest Germany and the Netherlands. The development of this region has been related to a technological revolution making possible suitable survey and exploitation methods and is a good example of the impact of high technology on resource recognition and exploitation. The rate of discovery of deposits and the pace of exploitation suggests that this will blossom into a major petroleum-producing region to remain significant – in fact, vital to the United Kingdom – until well into the next century. With conservative estimates suggesting reserves sufficient to make Britain ultimately self-sufficient in certain products, the North Sea Space had undoubtedly become even by 1975 a European energy focus. Forecasting the extrapolated development of the fields depends on a clutch of imponderables related to the continuation of additional worthwhile finds, but whether a find is 'commercial' rests ultimately upon world price levels in relation to exploration and exploitation costs. The North Sea basin also contains vast known reserves of coal, which high technology of the type used in oil development could make it possible to work.

North Sea oil has attracted a wide range of so-called 'oil-related' industries and considerable development of new supply bases in small harbours on Scotland's east coast and in Norway. Although a formidable refining capacity exists in the southern North Sea and Channel ports (especially Rotterdam-Europoort), there are proposals for new refineries around the northern basin (for example, in the Shetland Islands and the Cromarty Firth), with related gas processing (St Fergus–Peterhead). However, existing industrial areas (for example, Teesport and Emden) will also benefit, whereas by the 1980s a beginning to a considerable petrochemicals industry (for example, in Fife) could have been made. Quite apart from the massive dry dock and constructional facilities developed in southern Norway and in Scotland for rig and platform building, a mass of smaller industries have been crowding near to the oilfields. Deep water and plentiful room to maneouvre for large structures or even tankers in the northern basin compared with the narrow and shallow southern basin with its navigational hazards adds further weight to the possibilities of a wider development of the North Sea Space. If the 'Japanese connection' in the concept of large coastal steelworks with bulk carriers supplying raw materials and distributing finished products is pursued, many useful and advantageously located sites exist – a handful of vast iron and steelworks serving all Western Europe could be a reality early next century. German proposals for the development of the Elbe estuary and British developments on Teesside already point the likely pattern, though the choice of sites must be carefully made, because infrastructure costs can be extremely high, as found when a large integrated iron and steel complex was proposed on the Rotterdam-Europoort Maasvlakte some 30 km from existing services and the main residential areas for labour.

The Waters round Europe

The new role of the continental shelf and peripheral seas, besides the United Kingdom's closer involvement in the European Economic Community, sets a question mark over the role of the British Isles in Europe of the twenty-first century, especially as the Atlantic trading sphere develops. The bulk ocean carrier and the drift of heavy industry to coastal sites particularly favour Britain, for there are good deep-water locations suitable for mammoth coastal industries. Britain's harbours, with ready access to deep and open water, can also provide useful break-of-bulk points, where large ships can unload into smaller vessels for distribution to shallow water ports, or for distribution by land. The possibilities opened by developments of this type were well portrayed in the imaginatively conceived but unfortunately neglected *Oceanspan* project for Central Scotland in the early 1970s. The integration of British industry into that of the Greater Rhineland and elsewhere in the continental mainland in general will certainly demand better transport links by the year 2000 – the present ro-ro type ferries, even if greatly increased in size and number, will by then probably be unable to deal with the traffic and at least one Channel Tunnel will have become essential. One of the major growth axes in traffic across the North Sea will be between the *Eurodelta* and Humberside, with an enhanced arterial industrial belt between the Mersey–Dee estuaries and the Humber ports along the general axis of the M62 motorway, linked westwards to Ireland. From exploration already carried out, it looks as though by the end of the century an oil and gas industry could have developed in shelf waters along the Atlantic coast of Britain, most probably in the so-called Celtic Sea, that would be of the greatest importance to Wales and to Ireland.

The French Bridgeland

A number of economic forecasts in the early 1970s saw France as one of the richest and most advanced countries by the year 2000 and a quiet and steady social and economic revolution has been taking place. France will always gain from its role as a bridgeland – between Atlantic, Mediterranean and interior Europe – but a new attraction for bulk ocean transport terminals has enhanced the value of its Atlantic coast, while the importance of the Rhône corridor from the Mediterranean to the Greater Rhineland will doubtless increase. The decline of traditional heavy industries unable to meet rising demands of scale economies will reduce the former dominance of the north and the continued general diffusion of new industries could make 'le désert français' blossom. The already difficult energy supply problem is unlikely to moderate – France is therefore likely to attract most readily low energy industries, but as food supplies will become increasingly critical, the maintained strength of French agriculture could benefit the country even more.

The Twenty-first Century Eastern Europe

The unity forced on Eastern Europe by Soviet hegemony has created a situation not experienced before in history. Little major change in the broader political relationships can be expected even into the next century, but the grouping of the countries into more developed and sophisticated economies and a group of less developed ones with large agricultural sectors will tend to weaken over time as the latter move forward. The early postwar emphasis on self-sufficiency, and *Comecon*'s insistence on respect of national sovereignty, retarded the emergence of new regional shapes in contrast to the integrationist policies of the European Community, though by the mid-1970s some switch in emphasis in *Comecon* had become apparent.

In the circumstances, it is perhaps not surprising that no obvious replacement for the vanished *Mitteleuropa* appeared. Although there has been no comparable development to the Greater Rhineland, nevertheless during the 1970s an embryonic *Montandreieck* began to appear between the three most advanced economies – the G.D.R., Poland and Czechoslovakia – as greater co-operation, linkage and integration were officially encouraged. This will be one of the most powerful and important industrial nodes in the whole socialist *bloc* and has already shown tendencies to become the energy focus of Eastern Europe, as the emerging pattern of pipelines and electricity grids reflects. A powerful lobby within *Comecon* has put this core in a strong position to retain a greater part of the heavy industrial development, particularly in iron and steel and chemicals, for the future. Its links with the remaining *Comecon* members bear remarkable semblance to the prewar pattern of the now defunct *Mitteleuropa*. The long-term prospects are good, for its relatively sophisticated industrial structure forms a focus of high technology outside the Soviet Union, while within it the first multi- or trans-national industrial undertakings are being formed.

Danubia in the Year 2000

Prewar studies usually identified Danubia as a key region and certainly during the latter nineteenth century, owing to Habsburg influence, it had developed a considerable unity, though this was much disrupted in the economic nationalism of the interwar years. The Danube basin has remained divided between the Western-oriented Germany and Austria, the *Comecon* members (Hungary, Czechoslovakia, Rumania and Bulgaria) and the fence-sitting Yugoslavia, a political division that has kept the level of inter-state trade fairly low in contrast to the great arterial importance enjoyed by the Rhine in the integrationist European Community. To make the Danube a truly effective international waterway, not only must substantial navigation hazards be eliminated, but a much higher level of trade between riparian members stimulated to generate the requisite traffic. All riparian states could benefit from some overall plan for the basin, such as that suggested by the American geographer, George Kish, in his paper 'TVA on the Danube?'

published in the early postwar years.[3] Important developments have, however, taken place – the largest has been the joint Rumanian–Yugoslav
hydroelectric barrage at the Iron Gates, though an equally important
Rumanian project to build a canal across the Dobruja between Cernavoda
and Constanţa to avoid the difficult delta was vetoed by the Russians. But any
semblance of an overall plan is still absent from *Comecon*'s deliberations,
though Soviet ideas of the early 1950s for a complete high capacity inland
waterway 'ring' in Eastern Europe made the Danube a vital link in the system.
By the year 2000 a freer interchange of goods and further industrial growth
could make the Danube heavily used, and with access to the main bulk ocean
freight ports via the new canals, plus the use of modern techniques of river
navigation, it might become as attractive a heavy industrial axis as the Rhine.

Where have the Balkans gone?

In the early postwar years, common interests brought an attempt by Bulgaria,
Rumania and Yugoslavia to form a distinctive political association of Balkan
countries. Although this was thwarted by the Soviet Union because of its
political undertones, Soviet planners have nevertheless recognised the desir-
ability of creating some broad Lower Danubian or Balkan 'territorial production complex' covering at least Rumania and Bulgaria and having
linkages (for more than geographical reasons) with Soviet Moldavia and the
Ukraine. Although the Balkans in the one-time sense of a collection of small
antagonistic states has gone, the Balkan countries enjoy a promising future at
least through their considerable endowment with natural resources, whereas
their high rate of economic growth in the second half of the twentieth century,
if continued, could carry them to a position of strength by the year 2000.

The Mediterranean and Eurafrica

A macro-region likely to undergo great change in the twenty-first century is
the Mediterranean basin, where the concept of Eurafrica will possibly
emerge. The political–geographical ties that linked North Africa across the
Sahara and along the Nile have faded with the decline of the British, French
and Italian imperia, and in place there has been the rise in the Mediterranean
basin of an interface between American and Russian interests. Particularly
important has been the new sense of identity among the Islamic states of
North Africa, whereas on the European shore there is a divided identity between the socialist and non-socialist states. The development of oil, natural
gas and other mineral resources in North Africa in the 1960s generated a
rapid rise in wealth and a brisk exchange with the European Mediterranean
littoral that is likely to become increasingly significant by early next century,
unless there is some radical change in the pattern of energy use in Europe.
Travellers in North Africa sense a new vitality and a new direction that puts
some parts near 'take-off'.

As demographic change in the more developed European industrial economies slows population growth and creates problems of labour supply, the
more buoyant pattern in Southern Europe and North Africa will attract new

industries, particularly to reserves of under-utilised rural labour, a trend already seen in Italy and most recently in Iberia. If we accept that levels of affluence in Continental Europe will continue to rise into the next century (an assumption taken for granted in the mid-1960s but looking less certain by the mid-1970s), then the Mediterranean basin will continue to be the recreational playground for the remainder of Europe, provided adequate measures are taken to safeguard the environment, which has already begun to show signs of overloading. These roles will be enhanced if the Mediterranean can again establish its function as a major routeway from the Atlantic to the Indian Ocean through an adequate enlargement of the Suez Canal and a guarantee of uninterrupted passage, while this transit function will be emphasised by the growth of the economy of the Black Sea basin as well as by the Levant and Asia Minor.

Europe in the World of the Year 2000

A question vital to Europeans is whether Europe is likely to maintain its social and economic position in the world into next century and what changes may be expected. An answer is beset by many imponderables. If we map the world distribution of industrial and commercial phenomena, there appear to be five major agglomerations, all within narrow latitudinal limits in the northern hemisphere, and this belt of concentration is reflected in other distribution maps such as those of transport, where systems are principally focused between and towards these nodes. In this belt, however, no one centre at present overshadows the rest, nor does there seem much likelihood of this happening. Although North America may appear to stand in some respects by itself, if we were able to plot its interlinkages with Western Europe and Japan we should see these to be remarkably strong. The nodes of European Russia and Eastern Europe have, until recently, remained relatively isolated from the main nodes of Western Europe and Japan on their flanks, though closer ties now seem likely towards the end of the century. If we were to summate Western Europe (with its rising high technology industries) and Eastern Europe (with some remarkable growth rates), Europe as a whole would stand out as a particularly formidable agglomeration, even though the separate national economies are mostly of modest dimensions. Clearly the achievement of a high level of unity and integration within Europe is an important element for its future welfare, if not survival, in a highly competitive world. Although the 'European Age' has passed, Europe has lost little of its political–geographical significance in the world, demonstrated by the continuing major presence and even confrontation of the superpowers in Europe, despite their preoccupations elsewhere. Yet one is tempted to wonder whether this inability to find *détente* is not in some measure also a reluctance to disengage in Europe, because any withdrawal would leave a vacuum (even partial) to be quickly filled by a new sense of European oneness and the emergence of Europe as the much debated Third Force. A summated Europe has a very redoubtable potential.

Nevertheless, Europe is slowly losing its share of world population as its rate of growth slows while that of the lesser developed areas remains rapid. The marked ageing of the European population produces an inherent danger towards the year 2000 of a difficult social balance, while a drift towards 'zero growth' may not produce sufficient agile young minds to keep Europe's high technology research and development moving at an appropriate pace. An ageing population, if living standards are to be maintained, presupposes a rising ability to create wealth among the declining proportion of economically and demographically active age groups. There is, however, much opportunity to solve Europe's internal population imbalances by internal migration, a proposition made easier if nationalism and ethnic prejudices can be weakened.

Europe of the twenty-first century will be a corner of the great triangle of continents formed with Africa and Asia. Its contacts and involvement are likely to be greatest, however, in Asia, especially as the Eurasian landbridge between the Pacific and Atlantic spheres develops. Europe, like Japan, is well placed to take full advantage of low unit cost bulk oceanic freights, but the long-term significance of the, as yet, largely untapped resources of interior Asia will be formidable. Europe possesses countries wealthy and technologically advanced enough to have a major investment potential for interior Asian development, especially if the general upward trend in world commodity prices continues, so that even quite remote and inaccessible interior continental resources will have an attractiveness no longer deterred by costs. The high technology hearths of Europe and Japan will be well located to supply the rising demands, both quantitatively and qualitatively, inside the Soviet Union and in the more advanced Asian and African countries, while even mass production industries will benefit from the immense long-term market potential of Eurasia and later Africa, though the ultimate immensity of this market had hardly begun to appear by the mid-1970s. Although per capita purchasing power and consumption in China may remain low, the total Chinese market could be very large for capital goods. On the other hand, the long-term significance of the Middle Eastern markets may weaken as oil resources of other parts of the globe are developed or other forms of energy displace the present predominance of oil. If performance in the mid- and late twentieth century is representative, the Third World of Africa and South America has not yet got to the position of take-off already demonstrated by the socialist *bloc* countries. Yet it would be unwise for Europe to neglect its contacts with the Third World, another great potential source of raw materials and ultimately a great market.

Any idea that Europe has become a backwater no longer in the mainstream of world affairs and influence is surely fallacious. If the 'European Age' has passed in the sense of the great European imperia, the new move towards integration and unity will doubtless make Europe in every way equal to the perhaps misnamed 'superpowers', the United States and the Soviet Union. But in the world of the twenty-first century, it will be even more difficult than in our present century to consider any continent independently of the remainder: perhaps in terms of human geography we are quickly moving towards a conceptual *Pangaea* – the name given to the 'supercontinent', the overwhelming landmass on the earth's surface, that began to break up in the continental masses of our modern world about 200 million years ago.

References

1. See also W. Kennet (Ed.), *The Futures of Europe*, Cambridge University Press, London (1976), and A. Buchan (Ed.), *Europe's Futures–Europe's Choices – Models of Western Europe in the 1970s*, Chatto and Windus, London (1969).
2. The term 'post-industrial' was evolved by the American sociologist Daniel Bell in his *The Coming of Post-industrial Society*, Basic Books, New York (1973).
3. See G. Kish, 'TVA on the Danube?', *Geogr. Rev.*, **37** (1947), 274–302.

Bibliography

This bibliography contains only important general works and a few key articles. A very extensive range of regional geographies and specialised articles has had to be omitted in the interests of space.

Bath, B. H. Slicher van. *Agrarian History of Western Europe A.D. 500–1850*, Edward Arnold, London (1963).

Beckinsale, R. and Beckinsale, M. *Southern Europe: The Mediterranean and Alpine Lands*, University of London Press, London (1975).

Clout, H. D. *Agriculture Studies in Contemporary Europe*, Macmillan, London (1971).

Clout, H. D. (Ed.). *Regional Development in Western Europe*, Wiley, London (1975).

Clout, H. D. *The Regional Problem in Western Europe*, Cambridge University Press, Cambridge (1976).

Cole, J. *Italy*, Chatto and Windus, London (1966).

Cvijić, J. *La Peninsule Balkanique*, Armand Colin, Paris (1918).

Dandelot, M. and Froment-Meurice, F. *France*, Presses Universitaires, Françaises, Paris (1975).

de la Mahotière, S. *Towards One Europe*, Penguin, Harmondsworth (1970).

Dickinson, R. E. *The West European City*, Routledge and Kegan Paul, London (1951).

Elkins, T. H. *Germany*, 2nd edn, Chatto and Windus, London (1969).

Fisher, W. B. and Bowen-Jones, H. *Spain*, Chatto and Windus, London (1958).

Fitzgerald, W. *The New Europe*, Methuen, London (1946).

Franklin, S. H. *The European Peasantry*, Methuen, London (1969).

Franklin, S. H. *'Rural Societies', Studies in Contemporary Europe*, Macmillan, London (1971).

Geipel, J. *The Europeans: An Ethnohistorical Survey*, Longmans, London (1969).

Hodges, M. (Ed.). *European Integration*, Penguin, Harmondsworth (1972).

Houston, J. M. *A Social Geography of Europe*, Duckworth, London (1953).

Houston, J. M. *The Western Mediterranean World*, Longmans, London (1964).

Jordan, T. G. *The European Culture Area: A Systematic Geography*, Harper and Row, London (1973).

Kosiński, L. A. *The Population of Europe: A Geographical Perspective*, Longmans, London (1970).

Lambert, A. M. *Making of the Dutch Landscape*, Seminar Press, London (1971).

Malmström, V. *Geography of Europe: A Regional Analysis*, Prentice-Hall, New Jersey (1971).

Manners, G. *et al. Regional Development in Britain*, Wiley, London (1972).

Marchant, E. C. (Ed.). *The Countries of Europe as seen by their Geographers*, Harrap, London (1970).

Mead, W. R. *An Economic Geography of the Scandinavian States and Finland*, University of London Press, London (1968).

Mellor, R. E. H. *Comecon: Challenge to the West*, Van Nostrand, New York (1971).

Mellor, R. E. H. *Eastern Europe: A Geography of the Comecon Countries*, Macmillan, London (1975).

Mellor, R. E. H. *The Two Germanies: A Modern Geography*, Harper & Row, London (1978).

Minshull, G. N. *The New Europe: An Economic Geography*, Hodder and Stoughton, London (1978).

Monkhouse, F. J. *The Countries of Northwest Europe*, Longman, London (1964).

Monkhouse, F. J. *A Regional Geography of Western Europe*, 4th edn, Longmans, London (1974).

Mutton, A. *Central Europe*, Longmans, London (1961).

Ogilvie, A. G. *Europe and its Borderlands*, Nelson, Edinburgh (1957).

Parker, G. *The Logic of Unity – An Economic Geography of the Common Market*, Longmans, London (1969).

Pinchemel, P. (Ed.) *Regional Planning – A European Problem*, Council of Europe, Strasbourg (1968).

Pinchemel, P. *France: A Geographical Survey*, Bell, London (1969).

Pinder, D. *'The Netherlands', Studies in Industrial Geography*, Dawson, London (1976).

Pounds, N. J. G. *Europe and the Soviet Union*, McGraw-Hill, New York (1966).

Riley, R. *'Belgium', Studies in Industrial Geography*, Dawson, London (1977).

Robinson, E. A. G. *Backward Areas in Advanced Countries*, Wiley, London (1969).

Salt, J. and Clout, H. D. *Migration in Postwar Europe*, Oxford University Press, Oxford (1976).

Scargill, D. I. (Ed.). *The Problem Regions of Europe*, vols. 1–18, Oxford University Press, Oxford (1972–77).

Schmitt, E. *et al.* 'Deutschland', *Harms Geographie*, Bd. 4, List Verlag, Munich (1976).

Sinnhuber, K. A. 'Central Europe–Mitteleuropa–Europe Centrale', *Trans. Inst. Brit Geographers*, 20 (1954), 15–40.

Smith, C. T. *An Historical Geography of Europe before 1800*, Longmans, London (1967).

Sømme, A. *The Geography of Norden*, Heinemann, London (1961).

Stamp, L. D. and Beaver, S. H. *The British Isles: A Geographic and Economic Survey*, Longmans, London (1971).

Thompson, I. B. *Modern France – A Social and Economic Geography*, Butterworths, London (1970).

Turnock, D. *Eastern Europe*, Dawson, London (1977).

Walker, D. S. *Italy*, Methuen, London (1967).

Index

Numbers in italic refer to figures and tables